T0328631

INTELLIGENT COORDINATED CONTROL OF COMPLEX UNCERTAIN SYSTEMS FOR POWER DISTRIBUTION NETWORK RELIABILITY

INTELLIGENT COORDINATED CONTROL OF COMPLEX UNCERTAIN SYSTEMS FOR POWER DISTRIBUTION NETWORK RELIABILITY

XIANGPING MENG

ZHAOYU PIAN

ELSEVIER

Amsterdam • Boston • Heidelberg • London • New York • Oxford
Paris • San Diego • San Francisco • Singapore • Sydney • Tokyo

Elsevier
Radarweg 29, PO Box 211, 1000 AE Amsterdam, Netherlands
The Boulevard, Langford Lane, Kidlington, Oxford OX5 1GB, UK
225 Wyman Street, Waltham, MA 02451, USA

Notices
Knowledge and best practice in this field are constantly changing. As new research and experience broaden our understanding, changes in research methods, professional practices, or medical treatment may become necessary.

Practitioners and researchers must always rely on their own experience and knowledge in evaluating and using any information, methods, compounds, or experiments described herein. In using such information or methods they should be mindful of their own safety and the safety of others, including parties for whom they have a professional responsibility.

To the fullest extent of the law, neither the Publisher nor the authors, contributors, or editors, assume any liability for any injury and/or damage to persons or property as a matter of products liability, negligence or otherwise, or from any use or operation of any methods, products, instructions, or ideas contained in the material herein.

Library of Congress Cataloging-in-Publication Data
A catalog record for this book is available from the Library of Congress

British Library Cataloguing-in-Publication Data
A catalogue record for this book is available from the British Library

ISBN: 978-0-12-849896-5

> For information on all Elsevier publications
> visit our website at http://store.elsevier.com/

Typeset by Thomson Digital
Printed and bound in the United States of America

Working together
to grow libraries in
developing countries

www.elsevier.com • www.bookaid.org

Publisher: Joe Hayton
Acquisition Editor: Simon Tian
Editorial Project Manager: Naomi Robertson
Production Project Manager: Melissa Read
Designer: Matthew Limbert

CONTENTS

CHAPTER 1

Introduction

Contents

1.1 BACKGROUND AND SIGNIFICANCE OF INTELLIGENT COORDINATED CONTROL OF COMPLEX UNCERTAINTY SYSTEM

In the natural world and human society, things take on a form of systematic presence. Anything can be considered as a system, and also as a subsystem of a factor associated with other things in a larger system. The complex system with uncertain information is called complex uncertainty system [1].

More and more complex uncertainty control systems are characterized by high complexity and internal distributivity, and most of them are in a dynamic open environment, which is difficult to be controlled by traditional accurate-model-based control method [2]. The rapid development of modern computer and communication technology provides a strong technological support for system modeling, optimization control, and decision-making. Therefore, it has become a hot spot in the field of control and computer to build a new generation control theory of complex uncertainty system with a unified and more general model [3].

In the complex uncertainty control system with both continuous and discrete dynamic features, the controller should be designed to solve the problem of applicability of decentralized control, as well as

X. Meng and Z. Pian: Intelligent Coordinated Control of Complex Uncertain Systems for Power
Distribution Network Reliability. http://dx.doi.org/10.1016/B978-0-12-849896-5.00001-5

the coordination between each decentralized controller so as to optimize the overall control effects. Due to the difficulties from the perspectives of theory and practice, the coordination problems between controllers have not been well solved for a long time. Therefore, the current control methods of large-scale complex uncertainty system are mostly based on centralized control type or ad-hoc technique. However, due to the complexity of calculation, excessive communication, and lack of measurability in practice, the centralized control of complex uncertainty system is often unapplicable [2].

At present, there are three influential schools studying complex uncertainty system: the European School, Chinese School, and American School, among which the Chinese School is represented by Qian Xuesen's theory of "Open Complex Giant System" (OCGS) [4], the American School is represented by the theory of "complex adaptive system," (CAS) of Santa Fe Institute (SFI).

The most relevant contributor for the European School is the Brussels School led by Prigogine, which has carried out the research on conditions far from equilibrium and theory of dissipative structure; followed by the Haken School, whose theoretical flag is synergetics; and also by Eigen's hypercycle theory and complexity study in England. The European School's study on complex system is of strong humanities and philosophical style.

The representative of the Chinese School is Qian Xuesen, who put forward the theory of OCGS and its research methodology, namely "hall for workshop of metasynthetic engineering from qualitative to quantitative" system, which emphasizes the importance of human ingenuity and experience of practical activities, claiming to fully play the flexibility of mental intelligence, as well as the power of computers in calculation and information processing. The problem of OCGS can be recognized and solved through combining the wisdom of millions of people in the world with that of the ancients, "summing-up to get intelligence."

The American School is most famous for its complex adaptive system of SFI, which represents an important direction of complexity study and systematic theory, and is highly effective in solving a large variety of complex system problems; It is considered as "a brand new view of the world standing for a new attitude and a new perspective of analyzing problems." The CAS theory based on the idea of computer modeling is an innovation with specific methods and also in methodology.

The complex uncertainty system is of distinctive practicability, which can be realized through computer simulation. With the development of computer science and distributed artificial intelligence, the study on intelligence control and control method has become a hot topic for cross disciplines such as computer science, control theory, and information science, and the intelligence control algorithm with self-learning ability has become a new research focus [5–7]. The intelligence control algorithm is mostly applied in the highly interdisciplinary research field, integrating the different areas such as economics, logic, ecology, social science, and philosophy, etc.

Intelligent control technology is applicable to solve the complicated and open distributed problems, which can be adopted on the occasions with the following one or more characteristics [2]: (1) Statistics, control, and resources are in a distributed environment, calling for coordination in operation; (2) Software modules in the system are both independent and communicative with each other, which can realize some functions or tasks through cooperation or competition; (3) In order to fulfill some functions or tasks through coordination needed in the actual system on different hardware platforms by different programming languages, original software should be packaged to increase communication function. Therefore, the intelligent control technology attracts strong interest of researchers in each field due to its applicability to highly open and loosely coupled complex system and wide application in the areas such as complex system control, distribution network security control, flexible manufacturing system, computer network and software system, power system, and traffic control system, etc. [2,6,7].

The complex uncertainty control system boasts all the features that enable it to solve problems with intelligent control technology, so in order to adapt to the control requirements of distributed complex uncertainty system, the system control structure is divided into agents with independent functions according to the characteristics of system functions and physical properties. Through the communication and interaction of the agents, the coordination of all the controllers can be realized so as to ensure an optimized overall system control effect. Therefore, applying the intelligent control technology into the coordination control of complex uncertainty system represents a development trend of control theory, bearing great significance for both theoretical and applicable researches on the complex uncertainty system.

The security control of the power system is designed mainly to prevent various catastrophic accidents of blackout in large area, of which the ultimate goal is to provide all the key information required by the system in time, estimate rapidly the vulnerability of the system, and provide an overall real-time intelligent control system covering a wide area with adaptive self healing, adaptive network reconstruction, and adaptive protection [8–11]. Distribution network is an important part of power system, as well as a giant system of complex uncertainty, which is a line of transmitting power from the distribution transformer to the power point, so as to supply power for each distribution substation of cities and various power loads [8]. The security control of the distribution network is also an important guarantee for national production and daily life. Therefore, this book takes the security control of the distribution network as the research subject and analyzes deeply the intelligent coordinated control problems of the complex uncertainty system.

1.2 BACKGROUND OF DISTRIBUTION NETWORK SECURITY CONTROL

Electric energy has gradually become the most important energy of the world. With its support, the industrial and agricultural production, society's normal functioning and people's daily life can be sustained. Due to the rapid development of the national economy and improvement of people's living standards, the demand for power has also been increased greatly, encouraging rapid advance of the power system in recent years. Nowadays, the power system has gradually developed into a large-scale complex system, stepping into an era of big power grids, large units, and high voltage. The distribution network in safety operation, as an important part of power system, imposes direct impact on people's living standards and national economic development. However, due to the existing weakness of the distribution network and interference from external environment, accidents occur frequently, which result in cascading failures in some serious cases, affecting the power supply of large areas over relatively much longer time. Large area collapse of the power grid and blackout accidents will not only cause huge economic losses and affect people's life, but also endanger social security and destroy normal economic order. During the operating process, a large and complicated network will expose some unexpected vulnerabilities, which has recently been highlighted worldwide for its safe and stable operation [12].

In terms of the current distribution network in China, with the increasingly expanding scale of the modern power distribution network system and the improving voltage level of power grid, the planning, operation, and control of distribution network have become extremely complicated. Although the robustness and reliability of power grid are ensured as much as possible during the design and construction process, there are still many potential interferences and danger in reality, which in specific conditions and environment may cause cascading failures resulting in systematic disaster. A number of disasters in different countries have fully proved this point. So how to analyze accurately the current and potential dangers in distribution network, make scientific assessment on and effective control of its vulnerability is very important for a safe and stable operation of the distribution network [13].

1.3 RELATIONSHIP BETWEEN THE SAFETY AND STABILITY OF DISTRIBUTION NETWORK AND VULNERABILITY OF THE SYSTEM

The safety of distribution network refers to the endurance capacity of the distribution network under sudden disturbances (such as a sudden short circuit or unexpected loss of systematic components), which is the ability of supplying electricity and electric energy continuously to users when the power system withstands a sudden disturbance under dynamic condition. Safety can be divided into steady state and transient state corresponding respectively to the steady state safety analysis and transient safety analysis. The evaluating system of steady state safety focuses on a group of potential accident collections, mainly the small disturbances such as branch outages and load changes to check if lines overloading or exceeding voltage limit occur; and the evaluating system of transient safety which targets a group of potential accident collections, usually big disturbances such as bus short-circuit, sudden disconnection of lines, and generators, and so on, to check if stability is lost [14].

The stability of distribution network means that the distribution network resumes a stable operation mode or reaches a new stable operation mode through its inherent ability and stability control devices after it is disturbed.

As the most basic conditions, safety and stability are essential to normal operation of distribution network, yet they are different from each other. Safety means that all the power equipment must operate within the limits of current, voltage, and frequency amplitude and time. The safety of distribution network usually refers to the ability to avoid widespread power interruption when it is disturbed by sudden faults. Therefore, the emphasis of studies on safety and stability is different, for stability emphasizes the ability of sustaining the system's operation when the power system is disturbed, whereas safety focuses on the possible consequences of distribution network caused by a series of potential faults, as well as the endurance capacity of comprehensive evaluation system for various types of potential faults [15,16].

From the overall, micro and systematic perspectives, the assessment of the vulnerability of the distribution network analyzes its potential weakness. The loopholes are not selected by experience, instead all the power consuming devices in the power grid are analyzed and ordered, so that the loopholes of the power system can be analyzed and grasped from systematic and global perspectives. Thus the understanding of loopholes in the power grid can not only help the analyzing and dispatching personnel locate the research priorities before safety analysis, but also bring them a quantitative understanding of the distribution network from systematic perspective [9].

Therefore, the vulnerability assessment, safety and stability analysis of the distribution network are closely related and highly complementary tasks. The vulnerability assessment provides the theoretical basis for the selection of loopholes, which is the research focus of safety and stability analysis of the power grid. Through grasping the weak points of the system from micro perspective, an overall quantitative understanding can be achieved, thus deepening the safety and stability analysis work of the power system.

1.4 RESEARCH STATUS OF POWER GRID SECURITY CONTROL

In recent years, studies of the vulnerability assessment of power grid have increased. The earliest such study was raised in 1974, when the power grid structure was in relatively small scale, the traditional control mode was adopted for the power grid and the vulnerability analysis of power grid failed to get enough attention. As the grid control was removed, the traditional control mode has been changed into modern power market. Due to the economic pressure, the constructions of some new devices were delayed, yet the power consumption continued growing, making the existing devices operating around the edges. As the grid structures become increasingly complicated, accidents happen frequently, resulting in more and more serious consequences. People began reconsidering the loopholes in the structure and operation conditions of the power grid. It was not until 1994 that A. A. Fouad and others put forward the conception of vulnerability [17], which was considered as a new structure of the power system's dynamic safety assessment with the energy function method as an analysis indicator.

The distribution network might become vulnerable due to various reasons, including internal and external sources. According to different vulnerable sources, different vulnerability assessment methods of the power grid are put forward, mainly including the time-domain simulation method, the energy function method, and the hybrid algorithm. From the perspective of different vulnerable sources, they can be classified into deterministic and nondeterministic assessment methods [18].

The time-domain simulation method requires to solve nonlinear algebraic differential equations related to the power grid, demanding a large amount of calculations, which is usually applied in offline analysis. The model employed is very simple, which can only be used in short-term vulnerability research of power grid. The energy function method can describe the systems directly through energy function. Although it needs relatively less calculation compared to time-domain simulation method, it is not accurate enough. The hybrid algorithm is the combination of the above two methods, achieving the critical energy through the system time-domain integral equation to calculate the approximate trajectory of the system. The advantages of the hybrid algorithm lie in the ability to determine the relative safety under specific operating conditions. Yet when high precision is required, it can't be realized due to the model limit of energy function method.

The deterministic assessment method aims to determine the safety level of the system through checking the changes of the system's stable state upon a number of extremely serious accidents and defining the technical indicators with direct physical meanings, such as sensitivity technology, energy margin, power flow, and direct method, and so on, as the scale of evaluating the vulnerability of the power grid. These indicators usually take the most serious accidents as the standard, without any consideration of the randomness of accidents and complexity of the power grid. So its conclusions tend to be conservative, yet can serve as reference data of vulnerability analysis of the actual power grid [19,20].

The nondeterministic assessment method is divided into the probability assessment method and the risk assessment method. The occurrence of power grid accidents is of uncertainty, which still follows certain rules, for example, the probability of grid disturbance is in line with the Poisson distribution. The probability assessment method assumes the probability of some disturbances in the grid and the spread probability of transmission lines, determining the vulnerability of power grid through probability analysis. The risk assessment method is often adopted in the vulnerability analysis of power grid. In general, risk refers to the possibility of occurring specific undesired accidents and the integration of their consequences. Possibility and severity are two characteristics of risk, and the risk assessment method is aimed to study the possibility of dangerous accidents and the seriousness of their consequences. For the power system, it is a kind of measure that does not expect to disturb the possibility leading to dangerous operation and the consequences under the current operation environment of power grid, and the vulnerability of the power grid will be evaluated by this measure [21–23].

With the rapid development of automation technology, the application of network technology and the improvement of information technology, the existing power system adopts usually the centralized control system structure, which requires the control center to collect data from multichannel far away. So when faults occur, the control center can hardly undertake timely and effective interference in a short time. Meanwhile, during the normal operation process of the system, a preventive measure is needed to find or isolate the invisible faults in the system, and an intelligent distributed system is needed to exert random and real-time control on the complicated power system. But for the vulnerability and safety analysis of power grid, according to the research methods used by domestic and overseas experts mentioned above, the results only target one line, one substation, or bus of the power grid while ignoring the mutual influence among the other related lines, substations, or buses, which affects the overall vulnerability assessment of power grid. The intelligent control algorithm provides an idea for the vulnerability analysis of the distributed system of the power grid, which is also a trend for the future power grid vulnerability and security analysis.

1.5 MAIN CONTENT OF THE BOOK

In recent years, due to the scale expansion of modern power system network and the enhancement of the power grid voltage level, the security control of power system has become a main focus of power operators. Based on the analysis and summarization of domestic and overseas research results of the intelligent coordinated control theory, the book studies the internal control mechanism of the complex uncertainty system by adopting the multidisciplinary combination method such as machine learning, quantum theory, artificial intelligence theory, cybernetics, modern communication mechanism, and distribution automation, etc. The book studies the application of intelligent coordinated control algorism and takes the security control of the distribution network as the research subject of the complex uncertainty system. The book is divided into eight chapters with research details as follows:

Chapter 1 introduces the research background and status of the current complex uncertainty system and the security control of the distribution network, illustrating and analyzing in details the internal connection between the security control of the distribution network and the complex uncertainty system.

Chapter 2 introduces mainly the theoretical basis of intelligent coordinated control algorithm employed in the research, and summarizes its internal operation mechanism. And merits and demerits of each intelligent control algorithm, laying a theoretical foundation for the subsequent chapters.

Chapter 3 introduces mainly such concepts as grid vulnerability and vulnerable source, summarizes several kinds of current vulnerability warning indicators, and analyzes the characteristics of each indicator or index system. Then the applicable objects of various indicators, and their merits and demerits are obtained through the comparison of the mentioned indicators. Since an important part of grid warning system is to determine the indicators, this chapter aims to provide reference for the grid warning system.

Whether the voltage of distribution network is stable or not, it needs to be judged by indexes, and different analysis methods of voltage stability correspond to different analysis indexes. Chapter 4 introduces the current analysis indexes of voltage stability, as well as the voltage index L. Based on the voltage index L of the existing two nodes system, the multinode system is extended to deduce the expression of the voltage index L of load node in the multinode system and to put forward the quick calculation method to get an accurate value of index L.

Chapter 5 proposes first a new multiagent collaborative learning algorithm based on quantum (briefly known as Q-MAS) to solve the curse of dimensionality problems of behavior and state, improving the learning speed of the multiagent. Then for the vulnerability assessment and control problems of the distribution network, the multiagent coordinated control theory is applied to the vulnerability analysis of the distribution network for designing the hierarchical control structure of the distribution network based on Q-MAS, which can realize overall coordination of vulnerability assessment of the distribution network.

The security control of the bus in distribution network is subject to the vulnerability of the bus connected to it. However, most of the existing evaluation algorithms ignore the vulnerable information among the buses in low-voltage assessment, and merely evaluate the bus independently, which reduces greatly the accuracy of the assessment. To solve this problem, Chapter 6 introduces the assessment of mutual influence among buses, uses the combination of the analytic hierarchy process and fuzzy comprehensive evaluation to quantify it, and gets the comprehensive low-voltage risk assessment index by calculating the coordinated effect coefficient of vulnerable information among buses.

Chapter 7 presents a kind of ant colony algorithm based on direction coordination, which aims to strengthen the pheromone on the trail by defining the pheromone of a new direction, which improves the convergence speed of the algorithm to enhance the wholeness of solutions, apply the improved algorithm into the reconstruction of distribution network, put forward the updating method of corresponding directional pheromone and ordinary pheromone, and provide corresponding strategy of trails selection, which gives an effective solution for the reconstruction of distribution network.

Under the background of the optimization of maintenance plan of the traditional generator set, Chapter 8 establishes the optimizing model of the new maintenance plan with economic and technological considerations. To solve the disadvantages of the low convergence speed of the ant colony algorithm and easy to fall into local optimum, the dynamic transformation of two parameters affecting updating methods of the pheromone of ant colony algorithm is undertaken through fuzzy control rules, which enables adaptive adjustment under different conditions during the searching process of ant colony, affecting convergence speed and search conditions. Then the improved algorithm is applied into the optimizing model of the maintenance plan mentioned above.

BIBLIOGRAPHY

[1] Z. Janan, Hybrid systems and applications, Nonlinear Anal. 65 (6) (2006) 1103–1105.
[2] R. Negenborn, B. De Schutter, Multi-agent control of large-scale hybrid systems. Research Report, Delft University of Technology, 2004.
[3] T.G. Lyubomir, Nonlinear hybrid control systems, Nonlinear Anal. Hybrid Syst. 1 (2) (2007) 139–140.
[4] X. Qian, R. Dai, J. Yu, A new scientific field – open complex giant systems and its methodology, Nature 13 (1) (1990) 3–10.
[5] E. Ido, E.R. Alvin, Multi-agent learning and the descriptive value of simple models, Artif. Intell. 171 (7) (2007) 423–428.
[6] B. Lucian, R. Babuska, D.S. Bart, A comprehensive survey of multi-agent reinforcement learning., IEEE Trans. Syst. Man Cybern. Appl. Rev. 38 (2) (2008) 156–172.
[7] R. Guo, Research and Application of Multi-Agent Reinforcement Learning, Central South University, Changsha, (2005).
[8] J. Xu, Analysis of Anti-accident Ability and Fast Restoration of Large Area Outages of Distribution Network, Xi'an University of Science and Technology, Xi'an, (2008).
[9] X. Wang, T. Zhu, P. Xiong, Vulnerability assessment and control of MAS-based power system, J. Power Syst. Autom. 15 (3) (2003) 20–23.
[10] Y. Hou, L. Lu, X. Xiong, Y. Wu, Application of quantum evolutionary algorithm in transmission network planning, Power Syst. Technol. 28 (17) (2004) 19–23.
[11] J. Zhang, Q. Liu, Synergetic protection of multi-agent-based power grid, High-voltage Technol. 33 (1) (2007) 74–77.
[12] P. Wang, D. Xu, X. Wang, R. He, Study on vulnerability evaluation and protection of power system under strong external force harassment, J. North China Electric Power Univ. 32 (1) (2005) 15–18.
[13] W. Chen, Q. Jiang, Y. Cao, Z. Han, Vulnerability assessment of complex power system based on risk theory, Power Syst. Technol. 29 (4) (2005) 12–17.
[14] Q. Liu, J. Yao, L. Mu, J. Tang, Study on safety and stability of power grid under power market environment, Electr. Power Autom. Equip 23 (11) (2003) 73–76.
[15] X. Bai, Y. Ni, Overview of dynamic safety analysis of power system, Power Syst. Technol. 28 (16) (2004) 14–20.
[16] R. Billinton, S. Aboreshaid, Voltage stability considerations in composite power system reliability evaluation, IEEE Trans. PWRS 13 (2) (1998) 655–660.
[17] A.A. Fouad, Q. Zhou, V. Vittal, System vulnerability as a concept to assess power system dynamic security, IEEE Trans. Power Syst. 9 (2) (1994) 1009–1015.

[18] Y. Fan, X. Wang, Q. Wang, New improvement of safety analysis of power system – vulnerability analysis, J. Wuhan Univ. 36 (2) (2003) 110–113.

[19] Z. Qin, J. Davidson, A.A. Fouad, Application of artificial neural networks in power system security and vulnerability assessment, IEEE Trans. Power Syst. 9 (1) (1994) 525–532.

[20] D. Jian, B. Xiaomin, Z. Wei, F. Zhu, L. Zaihua, L. Min, The improvement of the small world network model and its application research in bulk power system, Int. Conf. Power Syst. Technol. (2006) 1–5.

[21] D. C. Elizondo, J. D. L. Ree, A.G. Phadke, S. Horowitz. Hidden failures in protection systems and their impact on wide area disturbances. The Bradley Department of Electrical Engineering, Virginia Tech, Blacksburg, VA 24061: 710–714, 2000.

[22] D.C. Elizondo, J.D.L. Ree, Analysis of hidden failures of protection schemes in large interconnected power systems, Proc. IEEE 93 (5) (2005) 956–964.

[23] M. Wang, Study on hidden failure, vulnerability and adaptability of automatic devices of power relay, Power Syst. Autom. Equip. 25 (3) (2005) 1–5.

CHAPTER 2

Theoretical Basis for Intelligent Coordinated Control

Contents

In recent years, with the development of control theory and computer science, and distributed artificial intelligence, intelligent control algorithm has led many interdisciplinary researches, such as computer science, control theory, and information science, bringing a new direction and new method for complex uncertainty system control.

X. Meng and Z. Pian: Intelligent Coordinated Control of Complex Uncertain Systems for Power Distribution Network Reliability. http://dx.doi.org/10.1016/B978-0-12-849896-5.00002-7

2.1 MULTIAGENT SYSTEM

2.1.1 Basic Concepts of Agent and MAS

The research of distributed artificial intelligence (DAI), and popularization of distribution network environment, has promoted the development of agent theory and technology, especially the multiagent theory and technology, for it provides a brand new approach for the analysis, design, and implementation of distributed open system. Since the concept of agent was first put forward in "Thinking of the Society," published in 1986, intelligent agent technology has matured rapidly, concomitant with the development of computer science, enjoying a good prospect in many application fields. Agent is currently one of the most widely used terms in the fields of computer science and technology, it appears in many research areas, such as artificial intelligence, DAI, distributed computing, human-computer interaction, software engineering, virtual reality, system simulation and computer-aided team working; it is also employed in the development of the software system in many application fields, such as workflow and business process management, heterogeneous information system, air traffic control, e-commerce, games, physical therapy care, and so on. Experts and scholars in these research and application fields usually understand the concept of agent, and analyze its properties and characteristics according to their own research aims, and the characteristics of their application fields. So, the meanings of agent usually depend on the specific area of research and application. For example, in the research area of artificial intelligence, people tend to focus on the intelligent characteristics of the agent, and its problem-solving ability; in the research field of distributed computing, researchers are concerned about the ability of agents to cooperate and move; in the application field of intelligent decision, people consider autonomous, learning ability, adaptability and intelligent decision-making as the main characteristics of the agent; yet in the e-commerce field, they regard autonomous, consultation, and spontaneous working as the main features of the agent. Therefore, there's no definition of the concept of agent currently that can be accepted and recognized by experts, scholars, and users of various research and application fields.

When Wooldridge [1], a famous research scholar on agent theory, discussed agent with others, he put forward two definition methods, those of strong definition and weak definition: the weak definition refers to the agent with basic characteristics, such as autonomous, social ability, reactivity and motility, and so on; the strong definition refers to the agent that not only includes basic characteristics of the weak definition, but also has the features related to specific areas, such as mobility, communication ability, rationality, etc. They believe that the agent should be a hardware environment or a software system that must have the following features:

1. Autonomous: the agent has the intrinsic computing resources, and the behavior control mechanism, limited to itself, and it can keep working in the absence of other agent or human intervention. This is the most basic characteristic of an agent, as well as an important feature making it different from general object.

2. Reactivity: the agent can perceive its environment (physical world, human users, or other related agent) and react appropriately to the relevant events that occurred in the environment.

3. Proactivity: the agent can take active action according to its targets, and even take the initiative to make a new target, and can tend to achieve it, meaning that agent is a goal driven behavior entity.

4. Social ability: when the agent is in an environment composed of multiple agents, its behavior must abide by the social rule of the agent group, and agent can make use of the information and knowledge of other agent in order to realize flexible and complicated interactions, cooperation, and competition with other agent through certain communication language, and so on.

In some special application systems, the agent still has many other features, such as being adaptive, mobility, rationality, and so on. Based on the differences in the characteristics of application systems and design requirements of software, different agents in different systems may show different degrees of behavior flexibility. Some application systems emphasize real time responses to events that require a higher reactivity of agent in the system; some systems emphasize the autonomous and intelligent behavior of agent, and focus on the proactivity of agent. There are also some agents whose behavior might show reactivity and social ability without proactivity, or show proactivity and social ability without reactivity.

In the field of software engineering, an agent is generally defined as a behavior entity capable of performing an action, autonomously and flexibly, in order to meet the design objective when residing in a certain environment. This definition is made from the perspective of software engineering that, firstly, is abstract and aims at no particular or specific research or application field with certain universality; secondly, it is of a top level that is not limited to any detailed implementation technology or development platform, so it is universal, in some circumstances.

The agent does not exist in isolation, it is usually with other agents in certain environments in order to realize a group of functions jointly, to provide a set of services, and constitute a system. At the same time, in many circumstances, the ability and resources owed by a single agent are limited, demanding multiple agents to collaborate with each other in order to fulfill the common task. The system constituted by several independent and interactive agents is called multiagent system, MAS. A multiagent system contains several agents, each of which is independent, and has a group of functions to provide a set of services; but these agents are interlinked, and have not only the correlation of structures, but also the correlation of behaviors. Furthermore, agents need interactions with the environment, such as getting certain resources from the environment, so the multiagent environment may occur in alternating and overlapping manner (such as sharing some public resources), fact that means that a certain relationship (such as two agents sharing one rare resource, leading to a conflict relationship, and requiring negotiation) exists between them. In order to realize complicated cooperation, competition, and consultation, communication languages and interaction protocols based on agents of high level (such as KQML) are needed between different agents, and interaction and communication can be realized through an interactive medium (physical devices, such as a network).

2.1.2 Organizational Structure of Multiagent System

Based on the presence of management and service in an MAS system, the organizational structure of MAS can be divided into three

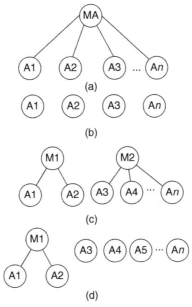

Figure 2.1 Organizational structure of multiagent system. (a) centralized type; (b) distributed type; (c) hybrid type 1; (d) hybrid type 2.

types: centralized type, distributed type, and hybrid type [1], as shown in Figure 2.1, among which the difference between centralized and distributed types is whether a center manager in charge of the centralized control of agent members is available or not.

The MAS with centralized structure integrates the closely related agents with shared desires into one group, and uses one management service institution to undertake the collaborative control of the group, under the premise that each member has certain autonomy. Multiagent groups can also form a group of senior level, and the collaboration of lower-level management institutions is subject to the senior management agent, with several such levels. Management service institutions and member agent have a certain relationship of controlling and being controlled.

Distributed MAS has no management service, yet it adopts the intermediary service agency to provide assistance and services for the collaboration of agent members, a situation that is not a management relationship.

However, a hybrid structure is characterized by a combination of distributed type and centralized type, with both a management service institution and an intermediary service agency. The decisive role of MAS organizational structure on the coordination mechanism is embodied in these two institutions that serve different functions in the collaboration.

The management service is responsible for the unified deployment and management of behavior, collaboration, task allocation, and resources sharing of all or part of the agent members, establishing the learning system and models of agent members to realize the safety inspection and control of the members' behavior and system [2]. The management agent and the member agent form a relationship of controlling and being controlled that, however, doesn't adopt a simple command mode, but rather a way of negotiation, so as to ensure the autonomy of agent members. A feasible management service should not only have overall objectives for the system, and knowledge about current environment, but it should also realize at least the following information: the location of all the agents within the jurisdiction; what kind of services and abilities do these agents provide and have; sharing information of resources within the jurisdiction, including varieties, quantity and usage, and so on.

Assume a management service institution as M, and all the processing application logic agents make up the set: Agent = $\left\{ \text{Agent}_1, \text{Agent}_2, \cdots, \text{Agent}_n \right\}$. In the collaborative process, if $\text{Agent}_i = (i = 1, 2, \cdots, n)$ can independently accomplish a task, consultation request to M is no longer needed, otherwise it has to offer a cooperation request. After receiving the request, when M finds out that this task can be finished by another one, or several agents, it can put forward a cooperation request to these agents, or tell them the information and leave it to them to negotiate. Upon receiving the information, the agent has the right to decide whether to accept the request of cooperation and gives feedback to M. This process will be repeated several times, until an agreement is reached. For complex tasks, M will decompose them and calculate the agent sets capable of fulfilling each subtask, and negotiate until an agreement is achieved.

By issuing, saving, and maintaining the information about the ability, position, and status of each agent member, the intermediary service agency undertakes the work of matching partners and service requests, and its relationship with agent members in the system is that of serving and being served.

The biggest difference between intermediary service organization and management service institution is that a management service institution has multiple agent members, and abundant knowledge about current environment of the system, whereas the intermediary service agency only provides services to agent members, without management of the system's sharing resources, lacking the ability of overall collaborative organization. This difference is determined by the scale of the system, and the intermediary service agency generally exists in large open MAS, and is generally not likely to have the overall knowledge within the system – such as the management service institution does.

2.1.3 Development Methods and Tools of Multiagent System

Due to the distributivity, complexity, and intelligence of multiagent system, the successful development of such a software system needs to be supported by new software engineering methods and tools. Because the abstract way of the agent and that of the objectives share certain similarities, relevant contents of the object-oriented technology can be used as references. Although, so far, no acknowledged development methods and tools of multiagent system are in shape, some research achievements made still bear reference significance.

(1) Agent oriented programming (AOP). Industrial application software usually consists of a large number of interactive parts with very complicated system, and this kind of complexity comes from the inherent complexity of the industrial system itself. The software programming aims to deal with the complexity more easily by providing means and methods on the structure and technology.

Agent oriented technology is suitable for the development of complicated software systems, with the reason of three aspects: (1) the agent-oriented analysis is an effective way for the space division of complex system problems; (2) the abstract way of agent oriented software is a natural way of modeling for complex systems; (3) the agent oriented philosophy can satisfy the management of complex systems on dynamic organizational relationships and structures.

The life cycle of agent oriented software development also includes demand analysis, system implementation and testing, and so on. The software engineering method of AOP is actually still in the beginning phase of research, without widely accepted complete theory and tools. Agent oriented programming methods are not yet mature, despite the emergence of a large number of application agents. In fact, most of the agent systems are not developed according to fixed software engineering methods; this is for the same reason that lots of software systems are developed before the emergence of software engineering.

(2) Development platform of multiagent system. Agent oriented programming is far from mature. So far, the development of agent projects all start from some basic software engineering practice. Through extracting the underlying services from the agent system, and forming a reusable application programming interface by abstraction, many agent development companies intend to simplify the development work of similar systems in the future, on the basis of which lots of development tools of multiagent system are formed, bearing different names, such as platform, environment, languages, framework, substructure, etc. At the present time, there are some famous MAS development tools, such as JATLite, Swarm, Jade, and so on.

2.2 REINFORCEMENT LEARNING OF MULTIAGENT SYSTEM

Russell and Wefald said, "learning is an important aspect of autonomy. A system may be called autonomous, which means that its behavior is determined by its current and past experience rather than the designer's input and experience. Agent is often designed for one class of environment, each of which is stored in Agent, consistent with the real environment known by the designers. When these assumptions are all true, the system operated on inherent assumptions can run successfully, so it is lack of flexibility. If enough reaction time is given, real autonomous system can operate successfully in any environment. In principle, the internal knowledge structure of the system may construct according to its own experience of the world, which can be used and modified by Agent during the problem solving process." The significance of learning ability on the agent can be seen from here.

2.2.1 Outline of Reinforcement Learning

In the field of machine learning, according to different feedbacks, learning techniques can be divided into supervised learning, unsupervised learning, and reinforcement learning [3], among which reinforcement learning is a special machine learning method that adapts to the environment with the input of feedback of the environment. Reinforcement learning refers to the learning process from the environmental state to the behavior mapping in order to maximize the value of accumulation reward, achieved from the environment by the system behavior. Different from supervised learning techniques that notify what kind of action should be adopted through positive and negative cases, this method aims to find the optimal behavioral strategy through trial and error methods [4].

Reinforcement learning is rooted in animal psychological research; the earliest case can be traced back to dog's reflection experiment conducted by Pavlov, and reinforcement learning was also named after this. At present, the reinforcement learning has developed into a multidisciplinary area, covering artificial intelligence, statistics, psychology, control engineering, operations research, neuroscience, artificial neural network, and genetic algorithm, and so on. Reinforcement learning is getting more and more attention because it can bring better design to an intelligent system operating under a dynamic environment. For example, a robot or robot agent system requires more autonomy in decision making in order to be effective under uncertain environments, as well as meeting the time limit. In these circumstances, learning is necessary to get skilled behavior, and it is only in such a case that reinforcement learning can show superiority over other learning methods.

Reinforcement learning gets behavior through trial and error methods, and the interaction of dynamic environment that is a process of mapping learning from environment to behavior, whose purpose is to maximize the scalar returns or the strengthened signal. Learners do not need to know which behavior to take, as most of the machine learning methods require, but they rather need to find out by experiment which behavior can bring maximum return. The choice of a behavior affects not only direct returns, but also the next environment, as well as all the returns followed. The search of delayed returns by trial and error is the most striking feature of reinforcement learning.

Motivating the agent through rewards and punishment requires no explanation about how the task succeeds. The key issues in reinforcement learning include establishing the basis of the field, learning from the delayed reinforcement, and setting up experience models to speed up the trade-off of exploration and exploitation, as well as generalized and hierarchical usage, and so on, through the Markov decision processes (MDP).

There are mainly two solutions to reinforcement learning. The first one is to find a behavior performing very well in the environment within the behavior space; that is the same with some other new search technologies that are applied to genetic algorithm and genetic design. The second solution is to use statistics techniques and dynamic planning methods in order to estimate the effectiveness of the behavior taken under a given world status. Which one is better and more effective depends on specific research application.

2.2.2 Principle of Reinforcement Learning

The environment faced by an intelligent system is usually dynamic, complex, and open requiring designers to subdivide the environment in the first place. Usually, the environment (problem) can be analyzed through five perspectives, listed in Table 2.1 [5].

In Table 2.1, episodic means that the knowledge learned by the intelligent system in each scenario is useful to the learning of the next scenario, just like a chess program facing the same opponent finds that the strategy learned at each game is helpful for the next

Table 2.1 Description of the environment

Point 1	Discrete state, continuous state	Point 4	Certainty, uncertainty
Point 2	State of total perceptibility, state of partial perceptibility	Point 5	Static state, dynamic state
Point 3	Episodic, nonepisodic		

game. On the contrary, a nonepisodic environment refers to the irrelevance of knowledge learned by the intelligent system in each scenario. Point 4 means that, in the environment of an intelligent system, if state transition is certain, the next unique state can be determined. Under uncertain environments, the next state relies on certain probability distribution. Furthermore, if the probability model of state transition is stable and unchangeable, it is called static environment, otherwise it is called dynamic environment. Obviously, the most complex kind of environment (or problem) is the dynamic environment under continuous state with, partly, perceptibility, nonepisodic and uncertainty.

Firstly, mathematical modeling shall be made on random, discrete state and discrete time, in reinforcement learning techniques. In practical application, the Markov model is the most commonly adopted. Table 2.2 shows several kinds of Markov models.

Based on the Markov decision process in Table 2.2, reinforcement learning can be simplified to the structure shown in Figure 2.2. In Figure 2.2, reinforcement learning system accepts input s of the environmental state, and, according to the internal reasoning mechanism, the corresponding action a is produced by the system. The environment is changed to new state s' under the influence of action a. The

Table 2.2 Several Markov models commonly used

Markov models	Whether intelligent system behavior controls the environmental state transfer		
		No	Yes
Whether the environment is partly perceptible	No	Markov chain	Markov decision process
	Yes	Hidden Markov model	Markov decision process with partial perception

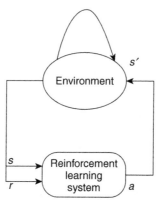

Figure 2.2 Structure of reinforcement learning.

system accepts the input of the new state of the environment, and gets the instantaneous reward feedback r about the system, given by environment at the same time. For a reinforcement learning system, its aim is to learn a behavior strategy $\pi : S \rightarrow A$, making the behavior chosen by the system get the maximum cumulative value of rewards from the environment. In other words, the system tries to maximize $\sum_{i=0}^{\infty} \gamma^i r_{t+i} (0 < \gamma \leq 1)$, of which γ is the discount factor, i is the number of iterations, and r_t is the rewards at time t. In a learning process, the basic principle of reinforcement learning technique is: if a certain action of the system gets positive reward of the environment, the tendency of producing this action of the system will be strengthened, and, if not, the tendency will become weaker, fact that is close to the principle of conditioned reflex in physiology.

If we assume the environment as a Markov model, sequential reinforcement learning problems can be modeled through the Markov decision process. First, the formal definition of the Markov decision process is put forward as follows.

The definition of $<S, A, R, P>$ quad contains a set S of environmental state, behavior set A of the system, reward function $R \ (S \times A \rightarrow R)$, and state transition function $P \ [S \times A \rightarrow PD \ (S)]$. Assume $R \ (s, a, s')$ as the instantaneous reward value achieved from transferring the environment state to s' by adopting action a at the state s. Assume $P \ (s, a, s')$ as the probability of transferring environment state to s' by adopting action a under the state s of the system.

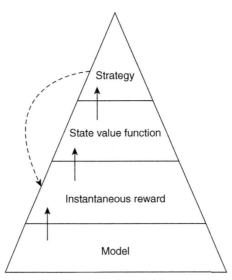

Figure 2.3 Four elements of reinforcement learning.

The nature of the Markov decision process is: the probability and reward value of current state transferring to the next state depends only on the current state and choice of action, instead of history status and action. Therefore, having known the knowledge of the environment model of probability function P and reward function R of state transition, dynamic programming technology can be used to solve the optimal policy. However, reinforcement learning focuses on how the system learning optimal behavior strategy functions P and R are unknown.

From Figure 2.3 we can see the relationship between four key elements of the reinforcement learning, that is, a pyramid structure. The environment faced by the system is determined by the environment model. Because the functions of model P and R functions are unknown, the system needs to rely on each trial and error to obtain the instantaneous reward to select the strategy. Because in the process of choosing a behavior strategy the uncertainty of the environment model and the long-term of goals should be taken into account, the value function (that is, the utility function of state) between strategy and instantaneous reward is built for the selection of strategy.

$$R_t = r_{t+1} + \gamma r_{t+2} + \gamma^2 r_{t+3} + \cdots = r_{t+1} + \gamma R_{t+1} \qquad (2.1)$$

$$V^{\pi}(s) = E_{\pi}\left\{R_t \middle| s_t = s\right\} = E_{\pi}\left\{r_{t+1} + \gamma V\left(s_{t+1}\right) \middle| s_t = s\right\}$$
$$= \sum_a \pi(s,a) \sum P_{ss'}^a \left[R_{ss'}^a + \gamma V^{\pi}(s')\right] \tag{2.2}$$

Firstly, construct a return function R_t through expression (2.1), in order to reflect the accumulative discount total of all the rewards received by the system after the state s_t, in a learning cycle guided by a certain strategy π. Due to the uncertain environment, R_t received in each learning cycle guided by some strategy π may be different. So, under strategy π, the value function of the system under state s is defined by expression (2.2) that reflects the sum of expected accumulative reward discounts that can be received by the system when following strategy π.

According to the Bellman optimal strategy formula, under optimal strategy $\pi^\star$, the value function of a system under state is defined by expression (2.3)

$$V^{\star}(s) = \max_{a \in A_{(s)}} E\left\{r_{t+1} + \gamma V^{\star}\left(s_{t+1}\right) \middle|_{s_t = s, a_t = a}\right\}$$
$$= \max_{a \in A_{(s)}} \sum P_{ss'}^a \left[R_{ss'}^a + \gamma V^{\star}(s')\right] \tag{2.3}$$

In the dynamic programming technology, under the premise of knowing the model knowledge of state transition probability function and reward function R, starting from the arbitrarily set strategy π_0, we can adopt a strategy iterative method to approach the optimal $V^\star$, $\pi^\star$, the dotted lines, as shown in Figure 2.2. Strategy iteration is shown in expression (2.4) and expression (2.5), in which k is the number of iterations.

$$\pi_k(s) = \arg\max_a \sum P_{ss'}^a \left[R_{ss'}^a + \gamma V^{\pi_{k-1}}(s')\right] \tag{2.4}$$

$$V^{\pi_k}(s) \leftarrow \sum_a \pi_{k-1}(s,a) \sum_{s'} P_{ss'}^a \left[R_{ss'} + \gamma V^{\pi_{k-1}}(s')\right] \tag{2.5}$$

The system cannot calculate the value function directly through expression (2.4) and (2.5) because in reinforcement learning, functions P and R are unknown. So, in practice we often use an approximation method to estimate the value function, and one of the main methods is that of Monte Carlo sampling, as shown in expression (2.6), in which R_t is the real accumulative reward discount value starting from the state s_t when the system adopts certain strategy π. Keep the strategy unchanged, and repeatedly use expression (2.6) in each learning cycle, until it is an approximation to expression (2.2).

$$V_{(s_t)} \leftarrow V_{(s_t)} + \alpha \left[R_t - V_{(s_t)} \right] \tag{2.6}$$

2.2.3 Multiagent Reinforcement Learning

Multiagent reinforcement learning is one of the most important research orientations of reinforcement learning study. In a multiagent system, the environment undergoes state transition under the joint action. For a single agent, it can only determine the actions of their own agent, so partial perception to a kind of behavior action is reflected, fact that produces another form of nonstandard Markov environment. Multiagent reinforcement learning mechanisms have been widely applied to various fields, such as games, mail routing, elevator group control system and robot design, etc.

Weiss divides the multiagent learning into three categories: multiplication, division, and interaction [6]. This kind of classification methods consider the multiagent system either as a computable learning agent, or that each agent has an independent reinforcement learning mechanism to speed up the learning process through appropriate interaction with other agents. Each agent has an independent learning mechanism, and the reinforcement-learning algorithm without interaction with other agents is called CIRL (concurrent isolated RL). CIRL algorithms can only be applied in the cooperative multiagent system that is better than the single agent reinforcement learning only in some environments. However, the reinforcement learning algorithm, with each agent having its own learning mechanism interacting with other agent, is called interactive RL. Different from single agent reinforcement learning that only considers temporal credit assignment problem, interactive RL mainly faces the structural credit assignment problem, that is, how to distribute the reward obtained by

Table 2.3 Three major multiagent reinforcement learning techniques

Learning techniques	Space of the problems	Main methods	Algorithm criteria
Cooperative multiagent reinforcement learning	Environment of distribution, isomorphism and cooperation	Exchange state Exchange experience Exchange strategy Exchange advice	Improve learning convergence speed
Multiagent reinforcement learning based on equilibrium solution	Isomorphic and heterogeneous, cooperative and competitive environment	Minimax Q Nash-Q CE-Q WoLF	Rationality and convergence
Multiagent reinforcement learning with optimal response	Heterogeneous and competitive environment	PHC IGA GIGA GIGA-WoLF	Convergence with no regrets

the whole system to each agent's behavior. Typical algorithms include ACE and AGE, and so on [7].

Weiss's discussion on multiagent reinforcement learning did not cover most of the research about current multiagent reinforcement learning. In fact, there are two perspectives of research methods in multiagent reinforcement learning: one is to start from the perspective of machine learning; the other is to analyze from the angle of multiagent system. In Table 2.3, three major multiagent reinforcement-learning techniques are given, followed by detailed analysis.

The first type of multiagent reinforcement learning is called cooperative multiagent reinforcement learning, and it emphasizes more how to use distributive reinforcement learning to improve the learning speed of reinforcement learning. In the early 1990s, Tan and others pointed out that the interaction (exchange information) in reinforcement learning should be one of the most effective methods [8]. Tan presented three ways to realize it: (1) to exchange the state information perceived from each agent perception; (2) to exchange agent learning experience episode, (also known as $< s, a, r, s' >$ experience); (3) to exchange the strategy or parameters in the process of learning. In 2004, Luis Nunes and others also proposed the fourth method: to exchange advice. Compared with single agent reinforcement learning, all of these methods can improve learning speed. The basic idea of cooperative multiagent reinforcement learning is to "interact and produce updated value function before agent choosing a behavior, while the selection of an action is based on new value function." However, Luis Nunes and other researchers stated clearly that the main problems of cooperative multiagent reinforcement learning is to solve "when and by which methods and reasons the information is exchanged and find out thoroughly under which condition it is absolutely beneficial to use this technology."

To answer this question and theoretically analyze the interaction of multiagent reinforcement learning, the researchers employed mathematical tool of the game theory to further analyze multiagent reinforcement learning. In gaming model, instantaneous rewards obtained by each agent depend on not only its own action, but also the actions of other agent at the same time. Therefore, each discrete state s of multiagent s reinforcement learning system can be formalized as a game g. So Markov decision model of reinforcement learning can be extended to Markov game model of multiagent system, which is called game-based multiagent reinforcement learning, or equilibrium-based multiagent reinforcement learning.

Analyzed using the Markov decision process, multiagent reinforcement learning can be understood as in n agent systems. Define the discrete state set S (that is the game set G), set A of action collection A_i, and combine the reward function $R_i : S \times A_1 \times \cdots \times A_n \to R$ with state transition function $P : S \times A_1 \times \cdots \times A_n \to P(S)$. Each agent aims at maximizing the expected discounted reward sum.

Multiagent game models basically can be divided into zerosum game, and broadsum game. Zerosum game stresses that the total reward obtained by multiagent from the environment is zero in any state of the system; meanwhile, broadsum game has no such requirement.

Obviously, if each state of the Markov decision process is formalized to the zerosum game model, the minimax Q algorithm can get the optimal solution that cannot be obtained in nonzero sum game models because it can reflect more about the conflicting nature of individual rationality and group rationality in multiagent systems.

Supported by the Nash equilibrium solution in game theory, Hu and the others designed the Nash-Q learning algorithm in order to get the optimal strategy in nonzero and Markov games [9]. Based on Hu's method, Littman and others considered two kinds of game models of zerosum and nonzero sum at the same time, and put forward the Friend-and-Foe-Q learning algorithm [10]; meanwhile Greenwald introduced a relevant equilibrium solution and designed the CE-Q algorithm to combine the Nash-Q learning algorithm and Friend-and-Foe-Q learning algorithm [11]. Among all multiagent reinforcement-learning algorithms based on equilibrium solutions, the algorithm must satisfy two properties – rationality, and convergence. The former shows that when other agent adopts a fixed strategy, the reinforcement learning algorithm based on equilibrium solutions should converge to the best response policy; the latter stresses that when all agents use a reinforcement learning algorithm based on the equilibrium solution, the algorithm must converge to stability policy without any oscillation phenomenon. We think the same: the basic idea of all multiagent reinforcement learning based on the equilibrium solution is that "when the agent chooses a behavior, it depends not only on its own value function, but also the value functions of other agents, which is a sort of equilibrium solution under all of the current agent value functions."

Different from any multiagent reinforcement learning algorithm based on the equilibrium solution, the best response multiagent reinforcement learning focuses on how to obtain the optimal strategy, no matter what kind of strategy is adopted by other agent. The two main criteria of algorithm design are the convergence criterion, and the no regret criterion. The convergence criterion is, in accordance with multiagent reinforcement learning algorithms, based on equilibrium solution. Whereas the no regret criterion refers to that: when the opponent policy is stable, the agent using the best response learning algorithm strategy gets rewards greater than, or equal to, the agent adopting any pure strategy.

The basic idea of best response multiagent reinforcement learning is to model the rival strategy, or optimize its own strategy. There are three main methods: (1) PHC, policy hill climbing algorithm that updates the strategy according to its own strategy history to maximize the reward; (2) optimize its own strategy according to the observation of the rival strategy [12]; (3) adjust the strategy by the gradient ascent method to increase its expected reward value. The typical algorithm includes IGA and its variant GIGA, WoLF, and GIGA-WoLF, etc. [8] We believe that the basic idea of best response multiagent reinforcement learning is that: "when the agent chooses a behavior, it doesn't only rely on its own value function, but also its historical strategy and the estimate of other agent strategy."

2.3 ANT COLONY ALGORITHM

2.3.1 Ants Foraging Behavior and Optimization

Ant colony optimization (ACO) algorithms are inspired by the foraging behavior of ants in the nature. In nature, some species of ants in searching for food will leave chemicals that can be smelled by others on the route, called pheromones [13,14]. By releasing pheromones, ants can mark the route they have walked, providing clues for other ants foraging for food. As time and the number of foraging ants increase, the concentration of pheromone in the environment will change, based on which ants can gradually find the shortest route between their nest and the food.

The most famous monitoring experiment on ants' foraging behavior is the "double bridge experiment" conducted by Deneubourg and his colleagues [13]. They used a double bridge to connect an ants' nest and the food source, studying the release of ants' pheromone and its influence on ants' foraging behavior. Assume two routes, a long one and a short one, connecting the nest to food. At first, ants chose both routes, but, after a while, all the ants chose the shorter one. In this experiment, at the initial stage, none of the two routes have pheromone, so ants had no reference for choosing, so there's no bias, and ants can only choose at random, so the probability of choosing one of the two routes is equal. However, the ants choosing the shorter route would arrive at the food source earlier, so they can get back earlier. They leave the pheromone on the shorter route next time, meaning that there is much more pheromone on the shorter route than on the longer one during the same period. Thus, as time passes by, accumulative concentrations of pheromone become greater and greater, compared with the shorter one, and the ants followed tend to choose the shorter route.

Inspired by this experiment, we can optimize the characteristics of real ants to make ants accurately remember the walked route that can release the pheromone according to certain values. The earliest ant colony algorithm is formed by using ants' foraging behavior in order to solve problems.

2.3.2 Artificial Ants and Real Ants

Artificial ants and real ants are adopted to distinguish the ant colony algorithm from ants in nature, with comparison between the two.

(1) Common points shared by artificial ant and real ant. As the ants in the ant colony algorithm are abstracted from the real ants in nature [15], the two share many similarities:

First, the two groups have communication mechanism. Whether in a real ant colony system or an artificial ant colony system, ants can release pheromones to change their environment, and communicate indirectly with other ants in the ant colony.

Second, they complete the same task through collaboration with other ants in the ant colony. Whether it is a real ant or an artificial one, the task is assigned to different ants in solving problems, and all the ants accomplish the same task through mutual cooperation.

Third, they all make use of the existing information to choose the route. Both real and artificial ants choose the route or node to be walked through according to the information of the environment they are in.

(2) The differences between artificial ants and real ants. In order to meet the demands of the algorithm, many characteristics that real ants don't have are added during the construction of artificial ants, mainly including:

1. Different from real ants, the artificial ants are in an environment that is a set of discrete state while solving problem, their movement is from one state to another one.
2. Real ants themselves can't record the route walked. But while setting artificial ants, make them capable of recording the routes they have walked through.

3. According to different problems, artificial ants can choose the updating method of pheromone, rather than update the pheromone each time they have moved a step, like real ants do.
4. According to the needs of the problem, artificial ants can add some functions which real ants don't have, such as adjusting the route selection plans dynamically, and overall updating, and so on.

2.3.3 Mathematical Models of Ant Colony Algorithm

The ant colony optimization algorithm is mainly used in solving combinatorial optimization problems. We adopt a triad (S, f, Ω) to show a combinatorial optimization problem, of which S represents the set of candidate solutions, f the objective function, Ω the set of constraints. Every candidate solution S has a corresponding objective function $f(s)$, and for all candidate solutions in the set S, the one that can satisfy the constraint conditions of Ω is called the feasible solution. The goal of the optimization algorithm is to find a global optimal solution s' among all of the feasible solutions.

In an ant colony algorithm, the ants build feasible solutions by moving on the building map $G_c = (C, L)$, of which $C = \{c_1, c_2, \cdots, c_n\}$ is the set of all the nodes in the map G_c, and L is the set of all the routes of nodes connecting the set C completely. Between the elements $c_i \in C$ and $l_{ij} \in L$ assume a pheromone value τ and a heuristic value η. In the ant colony algorithm, the pheromone value is updated by ants, in order to record the access of ants to the routes or nodes. Heuristic value η is determined by other factors besides ants, mainly including the existing information of problems to be solved, or other information during operations, generally used to represent the cost of new nodes or routes needing to be added in the solutions during the build process. In the process of marching, ants choose the next step in the map with certain probability, according to these values.

In the ant colony algorithm, all the movements between ants are parallel; meanwhile, the movements of each ant are independent from others. Although a single ant can also find the solution in a period of time after the search, during solving the problem, if a better solution is wanted, collaboration between ants is needed. Normally, this kind of interaction and the collaboration between the ants is realized through indirect communication by reading and writing pheromones.

In the ant colony algorithm, ant behavior k can be described as follows:

1. Ants look for the optimal solution through the access of each node and route on the building map $G_c = (C, L)$.
2. Each ant can save information on the path it has walked through. This information has the following functions: to establish a feasible solution, calculate heuristic η, evaluate current solution, and return according to the original road.
3. During the process of algorithm operation, each ant will be set up with an initial status and one (or more) termination status. The initial status is usually empty, or contains only the sequence of the initial node.
4. Under the state $x = <x_{r-1}, i>$, it can be determined whether ants meet the termination condition. If not, the ants will move on to the next node j, entering state $<x_r, j>$. If the ants do meet the termination condition, they finish this exploration and stop moving. During this process, when ants are choosing the state, they determine the next state through the recorded information or heuristic value η, and generally choose the next step from the set of feasible states, so in most cases they will not move to the nonviable state.
5. Ants choose the direction to move through state transition probability that refers to a function value determined by the following parameters in general: (1) the pheromone value and heuristic information value in the current environment, both of which generally refer to the pheromone value and heuristic information value on the node and route adjacent to its current position in the building map G_c; (2) the routes walked through by ants, that is the route information saved by ants currently; (3) constraints related to the issues.
6. When the ants add a node c_j to the current state, they will update the value of pheromone on the current node, or on the route related to this node.
7. When ants have completed a searching process and got a solution, they will backtrack according to the recorded information, and at the same time update the pheromone on the nodes along this route.

2.3.4 The Execution Process of Ant Colony Algorithm

The execution process of ant colony algorithm mainly can be divided into three parts:

1. The ants construct the feasible solution. Set the ants of the ant colony on each node of building map G_c, and the ants calculate the state transition probability p according to pheromone τ and heuristic information, and select the next moving direction. In general, when calculating the state transition probability, parameter $\alpha > 0$ and $\beta \geq 0$ is used, respectively, to determine the influence of τ and η value on p. In this way, a feasible solution of an optimization problem is gradually established through the ants moving to the adjacent nodes on the building map, fact which makes the ants in the ant colony visit asynchronously the adjacent state to the current one in parallel. During or after the process of building solutions, ants will analyze the information of established solutions in order to update the pheromone values of the next step.

2. Update the pheromone. The pheromone updating refers to the change of pheromone concentration, including increased and decreased pheromone concentration. After the visit of nodes or routes, or when this node or route is in optimal solution, ants will release pheromone on it, increasing the pheromone concentration on the node or route. When ants are choosing the node or route, they are more likely to choose the one with greater pheromone concentration; on the other hand, pheromone has an evaporation coefficient $\rho\,(0 \leq \rho \leq 1)$, namely the pheromone concentration will decrease at a certain ratio as time goes by, fact which will reduce the possibility of the algorithm converging to the nonoptimal solutions, expanding the searching scope of the algorithm.

3. Background operation. Background operation is mainly used for the global analysis and adjustment of the algorithm, and the process will collect global information, and adjust the algorithm accordingly, such as strengthening the pheromone on shorter routes, or adjusting the searching process by adding parameters, and so on. The process is optional, meaning that it can be chosen whether to perform the process or not, according to different problems.

The execution process of the ant colony algorithm can be represented by the following pseudo codes:

Start

While it doesn't meet the termination conditions

Manage ant's behavior

Update pheromone

Background operation

Finish

Among them, the three processes of managing the ant's behavior, updating pheromone, and background operation are collectively called behavior scheduling that can adjust the execution time and order, according to the specific requirements of problems during the problem solving process.

2.3.5 Basic Ant Colony Algorithm

The ant system (AS) is the earliest algorithm model proposed in the ACO algorithm [16]. In the AS, the ant k on the node i calculates the probability of reaching each node, according to the probability selection formula [expression (2.7)], and based on its result P_{ij} selects the next node j. Assume $tabu_k$ as the collection of all the nodes that ant k has visited currently, also known as the $tabu$ list of the ant, and add each node the ant has visited into the collection $tabu_k$.

$$P_{ij} = \frac{\tau_{ij}^{\alpha} \eta_{ij} \beta}{\sum_{h \in N_k} \tau_{ih}^{\alpha} \eta_{ih}^{\beta}} \tag{2.7}$$

Wherein, $N_k = \{C - tabu_k\}$ is the collection of all the nodes that ant k can visit currently, and when all the nodes in C are added into $tabu_k$, it means that all the nodes have been visited by the ant, with the ant coming back to the initial node. τ_{ij} is the amount of information along the route (i,j). η_{ij} is the heuristic function, with its expression in formula (2.8), the reciprocal of the distance d_{ij} between the two neighboring cities i, j. α is the information heuristic factor, and β is the expected heuristic factor, with α, β standing for the impact level on the algorithm of pheromone and heuristic information, respectively.

$$\eta_{ij}(t) = \frac{1}{d_{ij}} \tag{2.8}$$

In the AS algorithm, when the ant finished one adventure, it will update the pheromone along the walked path using such a method: make all the ants passing through the route release pheromone, and determine the value of pheromone to be updated, according to the following expression (2.9)

$$\tau_{ij} = (1-\rho)\tau_{ij} + \sum_{k=1}^{m}\Delta\tau_{ij}^{k} \tag{2.9}$$

Wherein, ρ is the coefficient of pheromone evaporation, a value ranging from [0,1], used to show the evaporation degree of pheromone as time goes by; $\Delta\tau_{ij}^{k}$ is the pheromone value released by the number k ant on the route (i,j) in this iteration process, whose value is determined according to expression (2.10).

$$\Delta\tau_{ij}^{k} = \begin{cases} F(k) & \text{if } (i,j) \text{ is an edge of ant } k\text{'s solutions} \\ 0 & \text{if } (i,j) \text{ isn't an edge of ant } k\text{'s solutions} \end{cases} \tag{2.10}$$

Wherein, $F(k)$ is the pheromone value left by ant k on the walked path (i,j), its size is often proportional to the reciprocal of the route's length solved by ant k, and multiply the reciprocal of the route's length by a constant Q, and Q is used to express the pheromone intensity. So, the shorter the route to be solved is, the bigger the value of $F(k)$ will be, and the more pheromones are released on this route by ants.

Dorigo once proposed three $\Delta\tau_{ij}^{k}$ algorithm models [17], namely, ant-cycle system, ant-density system, and ant-quantity system, whose major difference is the different value of $F(k)$.

In the ant-cycle system, $F(k)$ is related to the length L_k of the route solved by the number k ant in this iteration process, whose value is

$$F(k) = \frac{Q}{L_k} \tag{2.11}$$

In the ant–density system, $F(k)$ is related to the distance d_{ij} between city i, j, whose value is

$$F(k) = \frac{Q}{d_{ij}} \tag{2.12}$$

In the ant-quantity system, the value of $F(k)$ is a constant Q:

$$F(k) = Q \tag{2.13}$$

From expressions (2.12) and (2.13), we can see that ants use the local information in the environment in order to update the pheromone in the ant-density system and the ant-quantity system. In these models, once the ant has finished a selection of routes, it will update the pheromone of the chosen route that is also the local information. However, in ant-cycle systems, the algorithm uses the global information to update the pheromone, undertaken by the ant that has finished one journey, and the pheromone it has updated is the global information. The experiments show that, under the condition of using global information, the ant colony algorithm can make better performance to solve the problems, so the ant-cycle system is often adopted in order to update the pheromone in the algorithm when applying AS algorithm.

2.4 BP NEURAL NETWORK

The neural network has been applied widely in recent years, with a large number of varieties, mainly including back propagation (BP) neural networks [18], Hopfield neural networks, Boltzmann neural networks, and RBF neural networks, etc. Among those varieties, the BP network has been widely recognized by researchers because of its better function approximation ability, and effective training method. As such, this section also adopts the BP network.

The BP network was proposed and established by Rumelhart and other researchers in the early 1980s, and it was called multilayer feed-forward neural networks, the MFNN; it generally consists of an input layer, an output layer, and several hidden layers. Usually, a simpler BP network has only one hidden layer, or a network with three layers. The number of neurons of the BP network input layer and output layer is equal to the number of input and output of problems to be solved, while the number of neurons in the hidden layer is achieved under the guidance of an empirical formula. Connections are not allowed among neurons located on the same layer of the network, and neurons located on the adjacent layers are fully connected by weights that constitute a complete network.

2.4.1 Working Principle of BP Network

In order to describe the working process of a BP neural network, we assume the number of input layer neurons of a BP network as n, the number of hidden layer neurons R, the output layer neurons m. The connecting weights between the input layer and the hidden layer are $V = \left(v_1, v_2, \cdots, v_j, \cdots, v_R\right)$, with column vector v_j standing for the weight vector corresponding to the number j neuron in the hidden layer, b_j standing for the threshold value. The connecting weights between the hidden layer and the output layer are $W = \left(w_1, w_2, \cdots, w_k, \cdots, w_m\right)$, with the column vector w_k standing for the weight vector corresponding to the number k neuron in output layer, b_k standing for the threshold value. The input vector is $X = \left(x_1, x_2, \cdots, x_i, \cdots, x_n\right)^T$, the output vector of hidden layer $Y = \left(y_1, y_2, \cdots, y_j, \cdots, y_R\right)^T$, the output vector of output layer $O = \left(o_1, o_2, \cdots, o_k, \cdots, o_m\right)^T$, from which we can see that the input data is processed by connecting weights and functions from the input layer, and output data is finally achieved in the output layer; this means that the neural network is a nonlinear mapping from input to output.

The BP network is usually operated in two stages: the training stage, and working stage. During the training stage, sample data is used to train the network until the network parameters are adjusted to the optimal state, and then the network is saved. The working stage is to process the trained network, also known as simulation, during which each parameter of the BP network is constant, and the output value can be solved by inputting several groups of forecast data different from the training sample (Figure 2.4).

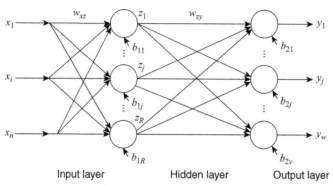

Figure 2.4 The structure of the three-layer BP neural network.

The training process of the BP network is divided in two stages, namely the forward transfer stage of input information, and the backward transfer stage of error; the detailed process is as follows.

1. Forward transfer process of information. The input data are transferred layer by layer from the input level to the output level, and the data reaching the output layer will be compared with the expected data, and if it does not reach the required error, the error will be transferred backward. The forward transfer process of data is as follows:

   For the hidden layer, there is:

   $$net_j = \sum_{i=1}^{n} v_{ij} x_i + b_j \left(j = 1, 2, \cdots, m \right)$$
   $$y_j = f\left(net_j\right)\left(j = 1, 2, \cdots, m\right)$$

   The input value of input layer is

   $$net_k = \sum_{j=1}^{m} w_{jk} c_j + b_{2m}$$

   The output value is

   $$o_k = f\left(net_k\right)$$

2. The backward transfer process of errors. The error signal is transferred layer by layer from the output layer to the input layer, during which the error is distributed to neurons of each layer,

continuously, producing the error signal of each-layer neurons, and the error signal can correct the weights and threshold values, whose process is as follows.

The error between the expected output value and the real output value of the number k sample is expressed like this:

$$E = \frac{1}{2}\sum_{k=1}^{l}\left(d_k - o_k\right)^2$$

Unfold the error to the hidden layer and the output layer, during which the error is:

$$E = \frac{1}{2}\sum_{k=1}^{l}\left\{d_k - f\left[\sum_{j=1}^{m} w_{jk} f\left(\sum_{i=1}^{n} v_{ij} x_i\right)\right]\right\}^2$$

In order to reduce the network error, the weights and threshold values are adjusted during the training process, and, at this moment, the adjustment of weights is proportional to the gradient descending of error, whose expression is:

$$\Delta w_{jk} = -\alpha \frac{\partial E}{\partial w_{jk}}\left(j = 1,2,\cdots,m; k = 1,2,\cdots,l\right)$$

$$\Delta v_{ij} = -\alpha \frac{\partial E}{\partial v_{ij}}\left(i = 1,2,\cdots,n; j = 1,2,\cdots,m\right)$$

Wherein, α is the learning rate, with minus standing for gradient descending. So, this algorithm is also called gradient descending algorithm and, thus, error E is connected with weights and threshold values that are adjusted through forward and backward transfer, until the weights meet the requirement.

2.4.2 Characteristics of BP Network

As a widely used network, the BP network has obvious advantages and disadvantages.

1. The advantages of the BP network include:
   a. The BP network can realize arbitrary complex nonlinear mapping between input data and output data, serving as a good tool to solve complex problems.

    b. In general, if the number of hidden layers of a network and neurons in each hidden layer is appropriate and enough, they can achieve the required mapping relationship.

    c. The speed of a trained BP network model could be fast, which generally can meet requirements.

    d. During the network training process, there are both inputs and expected outputs in the training samples. According to the size of errors and of target errors, adjust constantly the network parameter and finally achieve required errors through the comparison and training of errors between actual outputs and expected outputs. That is the network with self-learning ability under the learning style with mentors.

    e. After the network is trained, its predicative capability, also known as generalization ability, is relatively stronger, in general.

2. The BP network also has many disadvantages, as follows:

    a. Because there is no authoritative theory and computing method to determine the number of neurons in the hidden layer of the BP network, such determination is usually made through experiences or experiments, according to the problems to be solved, fact that not only costs more time, but also affects the performance of the network.

    b. Due to selecting the initial weights and thresholds at random, the BP network is easy to enter a local minimum value, away from the global optimal value, and the network models based on it cannot reach the expected target.

    c. To train a shaped network is generally time-consuming, after many times of training, so the speed of the BP network is slow.

    d. If the actual problem is too complicated, the network scale could be limited. In other words, the complex training process combined with additional learning samples will result in the decline of the network's performance.

    e. There are contradictions between the network training process and the forecasting process; this means that the forecasting effect may not necessarily be optimal with optimal network training effect. When the corresponding trend exceeds the optimal combining point, a better training result will lead to worse forecasting results; this is called overfitting. In other words, if taking account of too much samples while training the network, one may even ignore the innate characters of the samples.

2.4.3 BP Network of Increased Momentum Factor and Variable Learning Rate

According to the above disadvantages, the initial BP network is modified as follows:

1. The increased momentum factor. When adjusting the weights, the initial BP algorithm usually only considers the gradient descending direction of error at time t. The gradient direction before time is generally not taken into consideration, making the speed of oscillation and convergence of the BP network slower during the training process. In order to speed up the BP network training, the increased momentum factor η is adopted to adjust the weights formula, just like the BP network having a memory function, meaning that the momentum factor will take some of the last weights adjustment value and add it to the weights adjustment value of current subcycle, exerting a damping effect on weights adjustment at time, and improving the training speed with less oscillation frequency during the training process.

2. Adaptive learning rate. In the initial BP algorithm, learning rate α is constant. From the error surface, we can see that the flat area on the surface is relatively small, and the training speed of the BP network is slow; the precipitous area α on the surface is large, leading to a much larger adjustment value of the weights, which oscillates the training process and makes slower the training speed of the BP network. As such, the adaptive learning rate method is adopted in order to make the learning rate go up and down flexibly; this aims to improve the training speed of the BP network.

In this chapter, the BP network of three layers is adopted with only one hidden layer. Bearing in mind the low converging speed of the neural network, the BP network of three layers with variable learning rate and momentum factor is used. The scope of the learning rate is $0 < \alpha < 1$, of which, if α increases, the learning speed will rise, and changes of corresponding weights are also bigger. However, if α is too big, it will oscillate the learning process. So, we use a variable learning rate to get the ideal value through constant adjustment. In order to quicken the training process and avoid the training trapping in local optimization, the momentum factor η is introduced. If η is too big, it will oscillate the training process, so generally $0.7 < \eta < 0.9$ is adopted. The introduction of the variable learning rate and momentum factor improves both the convergence speed and the precision of the three-layer BP network.

2.5 PARTICLE SWARM OPTIMIZATION
2.5.1 General Introduction to Particle Swarm Optimization

Particle swarm optimization (PSO), originated from the analysis of behavior of birds catching food [19–21], was put forward by American scholars in the early 1990s. American scholars Kennedy and Eberhart found, during their analysis, that the flying birds often scattered, concentrated, or changed directions in an instant, adjusting their flight – fact, which is usually unexpected. After summarizing the rules, they found that the flying pace of the whole flock of birds would generally keep consistent, and a proper distance was maintained between each individual bird. Through analyzing constantly the behavior of other social animals, such as birds, fishes, ants, and so on, they concluded that, in the behavior rules of social animals, there has been an invisible information sharing platform for those seemingly unstructured and dispersed biological groups. Inspired by this, scholars simulated the behavior of birds constantly, and proposed the concept of particle swarm optimization.

Particle swarm optimization has become a better-developed optimization algorithm, in recent years. It searches the optimal solution through continuous iteration, and it finally employs the size of the value of objective function, or the function to be optimized (also known as the fitness function in the particle swarm), in order to evaluate the quality of the solution.

For the convenience of the study, birds are considered as particles of life without mass and volume in the algorithm. The algorithm initializes the position of each particle into the solution of problems to be optimized. In the movement process of the particle swarm, information is conveyed between each individual influencing the others, and a particle's moving state is influenced by the speed and direction of its colleagues, and of the whole particle swarm, so that each particle adjusts its own speed and direction according to the historical optimal positions of itself and its colleagues, and keeps flying and searching for the optimal position – the optimal solution. In the process of flying, particles update their position and direction according to their and external information; this has proved that the particle has the memory function, and particles with good positions and directions have the tendency to approach the optimal solution. As such, optimization is done through competition and cooperation between particles.

2.5.2 Basic Particle Swarm Optimization

(1) As an iterative optimization algorithm, particle swarm optimization was adopted to optimize continuous space problems at the early stage, and the description of particle swarm optimization was as follows:

When using particle swarm optimization to solve optimization problems, we can determine the states of all the particles by their flying directions, and speed, in the whole search space. Three vectors $\left(x_i, v_i, pbest_i\right)$ can be introduced to express a particle: x_i the current position of the particle; v_i the current speed of the particle; $pbest_i$ the best spatial position searched currently by the particle. During the process of PSO optimization, firstly initialize the solution of the optimization problem to a group of random solutions (a group of random particles); this also means to initialize the position of a particle to a random value, and then iterate gradually until an optimal solution is reached.

During the iteration, a particle updates its position and speed through two extrema: one is the optimal solution searched by the particle itself, called personal extremum *pbest*; the other is the one found by the whole group of particles, called group extremum *gbest*.

After finding these two optimal values, the particle will update its position and direction gradually, according to following formula:

$$v(k+1) = v(k) + c_1 \cdot rand(\ ) \cdot [pbest(k) - present(k)]$$
$$+ c_2 \cdot rand(\ ) \cdot [gbest(k) - present(k)] \tag{2.14}$$

$$present(k+1) = present(k) + v(k+1) \tag{2.15}$$

Wherein, $v(\ )$ is the speed of the particle; $present(\ )$ is the current position of particle; c_1, c_2 are learning factors, which are constants above zero; $rand(\ )$ is a random number between $[0,1]$. From formula (2.14), we can see that the updating of the particle's speed takes into account not only its historical speed, but also its personal and global best position, a fact which can approximate the particle to the optimal position. The speed involves a maximum value and a minimum value that are set artificially, and we assume the particle's speed to range between $\left[-v_{max}, v_{max}\right]$. During the process of particles gradually updating their speed, there are:

$$\text{If } v_{id} < -v_{max}, v_{id} = -v_{max} \tag{2.16}$$

$$\text{If } v_{id} > v_{max}, v_{id} = v_{ma} \tag{2.17}$$

(2) The procedure of basic particle swarm optimization is as follows:

1. Step 1. Initialize the parameters: initialize the position and speed of the particle to random numbers in the D-dimensional search space.
2. Step 2. Evaluate the particle's position: use a fitness function to evaluate each particle's position.
3. Step 3. Make a comparison between: (1) compare the fitness value of step 2 with the particle's personal best value *pbest*, and make the best value become the newest *pbest*; (2) compare the particle's fitness value with the global best value *gbest*, and the best one becomes *gbest*.
4. Step 4. Update the particle: according to the formula (2.14) and (2.15), update the particle's speed and position.

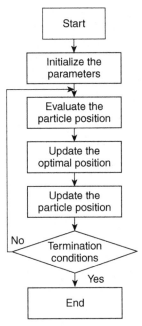

Figure 2.5 The procedure of basic particle swarm optimization.

5. Step 5. The termination conditions of iteration: circulate to step 2 until it satisfies the termination conditions, generally when the fitness value is optimal, or reaches the maximum iterations.

The procedure of basic particle swarm optimization is shown in Figure 2.5.

BIBLIOGRAPHY

[1] H. Li, Q. Wu, Review of multi-agent system, J. Tongji Univ. 31 (6) (2003) 728–732.

[2] W. Zhao, Y. Hou, Organizational structure and collaboration of multi-agent system, Comput. Eng. and Appl. (10) (2000) 59–61.

[3] G. Yang, Z. Zhihua, H. Jiazhou, et al., Study on multi-agent reinforcement learning model and algorithm based on Markov games, J. Comput. Res. Dev. 37 (3) (2000) 257–263.

[4] R.S. Sutton, A.G. Barto, Reinforcement Learning, MIT Press, Cambridge, MA, (1998).

[5] G. Weiss, Multiagent Systems: A Modern Approach to Distributed Artificial Intelligence, MIT Press, Cambridge, MA, (1999).

[6] G. Weiss, P. Dillenbourg, What is multi in multiagent learning?, in: P. Dillenbourg (Ed.), Collaborative Learning. Cognitive and Computational Approaches, Pergamon Press, Amsterdam, 1998, pp. 64–80.

[7] G. Weiss, Multiagent Systems: A Modern Approach to Distributed Artificial Intelligence, MIT Press, Cambridge, MA, (1999).

[8] M. Bowling, Convergence and no-regret in multiagent learning, in: L.K. Saul, Y. Weiss, L. Bottou (Eds.), Advances in Neural Information Processing Systems 17, MIT Press, Cambridge, MA, 2005.

[9] J. Hu, M.P. Wellman, Nash q-learning for general-sum stochastic games, J. Mach. Learn.Res. 4 (2003) 1039–1069.

[10] M. L. Littman. Fierend-or-foe q-learning in general-sum games. In: Proceedings of Eighteenth International Conference on Machine Learning, Williams College: Morgan Kaufman, 2001, 322–328.

[11] A. Greenwald, K. Hall, R. Serrano. Correlated-q learning. In: Proceedings of Twentieth International Conference on Machine Learning, Washington DC, 2003, 242–249.

[12] C. Claus, C. Boutilier. The dynamics of reinforcement learning in cooperative multiAgent system. In: Proceedings of the Fifteenth National/Tenth Conference on Artificial Intelligence/Innovative Applications on Artificial Intelligence, Madison, Wisconsin, United States: American Association for Artificial Intelligence, 1998, 746–752.

[13] J.L. Deneubourg, S. Aron, S. Goss, et al., The self-organizing exploratory pattern of the Argentine ant, J. Insect Behav. 3 (1990) 159–168.

[14] S. Goss, S. Aron, J.L. Deneubourg, et al., Self-organized shortcuts in the Argentine ant, Naturwissen-schaften 76 (1989) 579–581.

[15] M. Dorigo, G. Dicaro, L.M. Gambardella, Ant algorithms for discrete optimization, Artif. Life 5 (2) (1999) 137–172.

[16] M. Dorigo, V. Maniezzo, A. Colorni. The ant system: an autocatalytic optimizing process. Tech. Rep. 91-016 Revised, Dipartimento di Elettronica, Politecnico di Milano, Italy, 1991.

[17] M. Dorigo, V. Maniezzo, A. Colorni, The ant system: optimization by a colony of cooperating agents, IEEE Trans. Syst. Man Cybern. 26 (1) (1996) 29–41.

[18] G. Yang, D. Huailin, Study on Application of PSO-BP Neural Network on Credit Risk Assessment of Commercial Banks, Xiamen University, Xiamen, (2009).

[19] G. Chen, J. Jia, Qi. Han, Study on decreasing inertia weight strategy of particle swarm optimization, Acad. J. Xi'an Jiao Tong Univ. 40 (1) (2006) 53–61.

[20] A. El-Gallad, M. El-Hawary, A. Sallam, A. Kalas, Enhancing particle swarm optimizer via proper parameters selection., IEEE CCECE02 Proceedings, 2002Piscataway, NJ. 2.

[21] J. Kennedy, The particle swarm: social adaptation of knowledge, IEEE International Conference on Evolutionary Computation, 1997Indianapolis, IN.

CHAPTER 3

Power Grid Vulnerability Early-Warning Indicators

Contents

3.1 INTRODUCTION

With the rapid economic development and advanced modernization level, our dependence on electricity increases continuously, setting higher requirements on power grid security and stability. Over recent years, the power system has witnessed frequently occurring accidents, such as the "9.26" Hainan blackout in 2005 [1], blackouts in southern China due to snowstorms in 2008 [2], the "8.14" blackout in America and Canada in 2003, the "9.23" blackouts in Sweden and Denmark, the "9.28" blackout in Italy [3,4], and the "5.25" Moscow blackout in 2005. These mass blackouts caused inconvenience to production and to the lives of people, and even huge losses to the national economy. Therefore, prediction and forecast on the power grid's safety and stability, and the vulnerability analysis and early warning, is currently an important task for electricity operators.

X. Meng and Z. Pian: Intelligent Coordinated Control of Complex Uncertain Systems for Power Distribution Network Reliability. http://dx.doi.org/10.1016/B978-0-12-849896-5.00003-9

In the early-warning power grid security control process, it is an important link to determine vulnerability indicators. This chapter summarizes the existing vulnerability assessment indicators of power grids, and analyzes the characteristics of various indicators, providing valuable reference data for the power grid's safety operation and early warning.

3.2 POWER GRID VULNERABILITY AND INFLUENCE FACTORS

3.2.1 Power Grid Vulnerability

In recent years, the word vulnerability often appeared in relevant literatures of many fields, such as environment, ecology, computer network, and power system, and it is used to describe relevant systems and related components that are easy to be affected and destroyed, incapability of anti–interference and restoring initial state (the structure and function of itself). Words similar to vulnerability in meaning include sensitivity and fragility, but they have different meanings in different disciplines. System vulnerability is closely related to the system's safety level and its changing trend with changes of parameters. In this concept, an acceptable benchmark is set in terms of system safety indicators, and the trend of system state. After the system safety status is assessed, its safety level and changing trend can be determined. Whether the system is vulnerable or not depends on the comparison between its safety level and benchmark [5–7].

In the power grid, vulnerability is defined as a dangerous state when the power system faces potential mass blackout due to human intervention and other factors, such as information, computing, communication, internal components and protection control system, and so on, and the hidden trouble of this dangerous state or accident is exposed when the system failure occurs, represented by the system's capability to keep stable and normal power supply [8,9].

3.2.2 Influential Factors of Vulnerability

As power grids are becoming continuously complicated, with increased random disturbance, they may become even more vulnerable due to their own reasons, and to external disturbance. Besides unreasonable operation, topology changes can also lead to vulnerability of the power grid. In a word, factors resulting in vulnerability can be summarized as external factors and internal factors [10,11].

1. The external factors.
   a. Natural disasters and climate conditions: earthquakes, hurricanes, hail, thunderstorm, storms, floods, heat waves, forest fire, pollution flashover, and animal triggers.
   b. Human factors: installed error of control and protection system, incorrect operation by system operators, and sabotage, etc.
2. The internal factors.
   a. Faults in the main components: transformers, power lines, generators, etc.
   b. Control and protection system failure: hidden failures of the protection system, circuit breaker failure, interactions of the control, control system failure or protection misoperation, etc.
   c. Information/communication system failure: losing communication with EMS, incapability of automatic control and protection, error or congestion of the information system, external agent intrusion into the information/communication system, etc.
   d. Unstable power system, including voltage instability, dynamic and static instability, oscillation frequency instability, etc.

3.3 EARLY-WARNING INDICATORS OF POWER GRID SAFETY

The vulnerability early-warning indicator is used to describe the vulnerability level of the power grid that can reflect endurance capacity when accidents occur in the power grid. Since vulnerability indicators are different, in terms of different perspectives of power grid such as voltage, steady state, frequency and disturbances in various forms, specific research objects must be focused on when to determine vulnerability indicators; also, factors having a great impact on research objects should be taken as indicators, and certain principles must be followed in selecting the indicators. Here, the current vulnerability indicators will be summarized as references for the power grid early warning.

3.3.1 Risk Assessment Indicators [12–14]

On the basis of overcoming the demerits of deterministic and probabilistic assessment methods, a risk-based quantitative assessment method is put forward from two perspectives: probability, and impact of accidents. Indicators determined by this method are risk indicators. If the risk indicator value is big, it means that the corresponding system is more vulnerable, but if the value is small, it means that the system is relatively strong. The quantitative calculation formula of a risk indicator is:

$$R(C / X_t) = \sum_i P(E / X_t) \times S(C / E) \tag{3.1}$$

Wherein, i is the set of elements that operate normally in the assessment; X_t is the operating state before faults occurred; E is an uncertain accident; C is the result of the uncertain accidents; $P(E/X_t)$ is the probability of occurrence of E under X_t; $S(C/E)$ is the severity of consequence C caused by E; $R(C/X_t)$ is the risk index value.

Reference [12] mentions the risk indicator of overloading, low voltage, and loss of load. The risk indicator of low voltage targets each bus node that can reflect the possibility and damage level of voltage falling of each bus in the system that was caused by accident. The risk indicator of overloading focuses on each line that reflects the overloading conditions of lines after an accident. The system voltage instability risk indicator gives a risk value of instable system voltage from the perspective of the system, and finally calculates the overall risk indicator. The risk index is a kind of steady state vulnerability indicator, resulting in only a specific value that is inapplicable to online early warning and decision-making.

3.3.2 Early-Warning Indicator of Voltage Stability

Reference [15] proposes an early-warning model of voltage stability, based on an artificial immune power system, and lists a series of interrelated indicators that can reflect sensitively the voltage stability state, and problems of the power system for early-warning analysis, as shown in Figure 3.1. Afterwards, an artificial immune algorithm is employed to warn the node voltage early. This indicator system is relatively comprehensive, and worthy of reference.

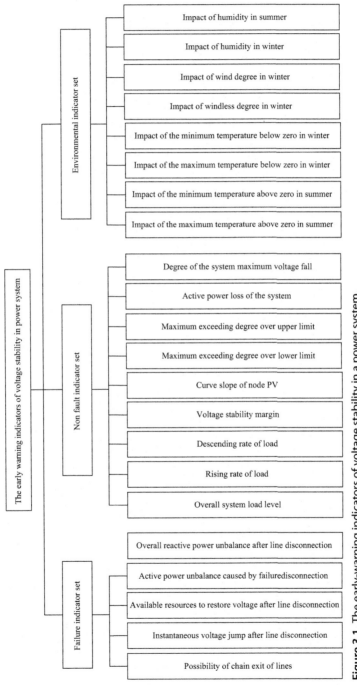

Figure 3.1 The early-warning indicators of voltage stability in a power system.

3.3.3 Safety Assessment Indicator of the Power Grid

Network safety is very complex, and it involves a wide scope. Describing it by using a single index will be inevitably affected by random error. Besides, due to the uncertainty of power grid operation, a large number of observation results are needed in order to grasp its internal rules. Multiple indicators can comprehensively reflect its overall situation and relationship that is more reliable for the assessment of the power grid. Reference [16] puts forward comprehensive safety evaluation indicators, as shown in Figure 3.2, based on which the calculation and analysis system of safety evaluation indicators is developed. Application in some power grids shows the rationality and practicability of this indicator system.

3.3.4 Grid Safety Evaluation Index System

At the current time, some literature is based on only one particular aspect or element of the power system, when evaluating its grid vulnerability, security, and so on. With the interconnection of the power grid, long-distance large-capacity power plants, and the application of new technologies and products in the power system, assessments on the running state of the power grid should be made from a macroscopic angle [17–19].

Reference [19] puts forward the power grid safety evaluation system that considers the power grid to be evaluated as a generalized node system, as a whole, as well as the influences of mains side, consumer side, and other connecting power grid, on the safety of the node system. It analyzes the main factors affecting the safety of the power grid, providing the method to analyze quantitatively the uncertainty factors, through three indicators: change, change rate, and sensitivity coefficient – all of which construct a three-layer indicator system of power grid safety assessment, as shown in Figure 3.3.

This three-layer indicator system reflects fully the main factors affecting grid security, and analyzes the influences on grid safety of uncertain factors; this is in line with the actual grid operating conditions with high operability, and has certain values for reference.

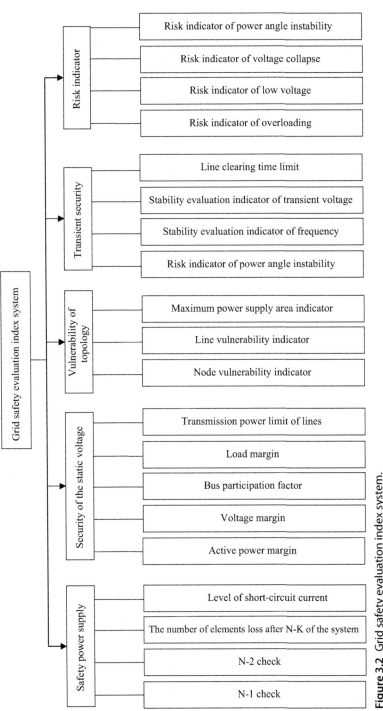

Figure 3.2 Grid safety evaluation index system.

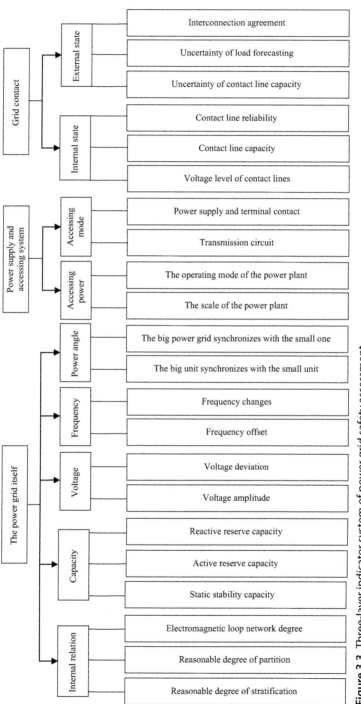

Figure 3.3 Three-layer indicator system of power grid safety assessment.

3.3.5 Frequency Vulnerability Indicator

With people's dependence on electricity being increased, and the power market competition intensified, requirements for power quality have become increasingly higher, and frequency is an important factor affecting power quality. When the frequency exceeds the permissible normal fluctuation range, it will not only damage the electronic equipment, but also impose a huge impact on the power system and power plants, bringing economic losses. Since the active power in the power grid and the frequency are closely related, the lack of active power in the power grid will decrease the frequency, or even cause the collapse of frequency at times when active power is in serious shortage. So, the following formula is used to measure the vulnerability indicator of frequency [18,19].

$$A = \frac{\partial f}{\partial P}$$

$$B = \frac{\Delta f}{\Delta P}$$

$$C = \begin{cases} \dfrac{49.8 - f}{49.8} & f \leq 49.8\,\text{Hz} \\ 0 & 49.8 \leq f \leq 50.2\,\text{Hz} \\ \dfrac{f - 50.2}{50.2} & f > 50.2\,\text{Hz} \end{cases}$$

$$D = ABC$$

Wherein, A is the variation of the power system frequency based on active power deficiency; B is the unit regulating frequency of the power system; C is the frequency range of the power grid with ± 0.2 Hz as allowable value in the power system; D is the vulnerability index of grid frequency.

Nodes with a greater value of B suffer more vulnerability, compared with other nodes in the power grid.

3.3.6 Early-Warning Indicator of Natural Disaster

For a long time, power grids have suffered from snow, lightning, thunderstorms, high winds, heavy rain, typhoons, fog and earthquakes, and other frequently occurring natural disasters, so the requirements for forecasting and early-warning information about natural disasters, such as grid scheduling, load forecasting, operation mode, and so on, are increasingly higher. However, the basis for early-warning work is to determine early-warning indicators and Figure 3.4 shows some reference indicators [20,21].

3.4 ADVANTAGES AND DISADVANTAGES OF EACH INDICATOR AND SCOPE OF APPLICATION

Analyze, by comparison, each indicator mentioned above, and find out its merits and demerits and applicable scope, as shown in Table 3.1.

The research on vulnerability problems of the power grid has achieved positive results, providing the theoretical basis for early warning and for the safety operation of the power grid. However, for the specific power grid, or a certain aspect of it, some indicators may be appropriate and some may not. So it is an important work to choose practical, operational, influential indicators based on the study object; it is only in this way that an intelligent, online and real-time power grid early warning can be achieved.

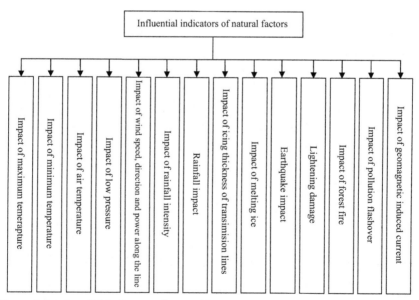

Figure 3.4 Influential indicators of natural factors.

Table 3.1 Comparison analysis of indicators

Indicator	Advantages	Disadvantages	Applicable object
Risk assessment indicator	Quantifies the possibility and seriousness of grid accidents, and gives a quantified risk indicator value on the whole grid, elements, or nodes	As a static value, it can only show the size of vulnerability of the power grid, elements, or nodes; lack of a comparison standard and less specific	Suitable for offline analysis on both the transmission and distribution network of the power grid, informing dispatchers of the weak areas and aspects of power grid in advance, playing a role in economic operation of the power grid
Early-warning indicator of voltage stability	Takes failure, nonfault and environmental factors into account, and analyzes in detail their impact on voltage stability	Without taking into account large disturbances, such as the impact on voltage stability after losing generator. Indicators are in large number with disunified dimension, lack of normalized analysis	Suitable for the transmission network, lacking modification when applied to the distribution network warning, especially the impact of distributed generator's access
Grid safety evaluation indicator	Considers several aspects such as static, transient, topology, safe power supply capacity, risk indicator; and subordinate indicators of each index are comprehensive, with clear physical meaning and high operability	Without consideration of the safety of the power generator, users and contact of network between each level	Suitable for transmission network without the supply side and consumer side; for distribution networks, no need to consider the transient; as topology is different, the checking of safe power supply capacity is also different

(Continued)

Table 3.1 Comparison analysis of indicators (*cont.*)

Indicator	Advantages	Disadvantages	Applicable object
Three-layer indicator of grid safety assessment	Considers the safety of power grid itself, power generator and network contact; and each indicator has clear physical meaning	Consideration for the safety of the power grid itself is not comprehensive enough, such as static stability and voltage stability, and so on	Applicable for transmission network
Early-warning indicator of natural disasters	Considers each natural disaster comprehensively	Indicators are in large number, with disunified dimension, lacking normalized analysis	Applicable for both transmission and distribution networks
Vulnerability indicator of frequency	Evaluates the vulnerability of grid frequency from the aspect of nodes	Lack of vulnerability assessment of frequencies between power grid and power generator, and each power generator	Applicable for both transmission and distribution networks

BIBLIOGRAPHY

[1] M. Zhang, Inspiration from black-start process of power grid after Hainan "9.26" mass blackout, East China Elect. Power 33 (11) (2005) 13–16.

[2] D. Shao, X. Yin, Q. Chen, et al. Influential analysis of blizzard disasters on southern china power grid in 2008, Power Syst. Technol. 33 (5) (2009) 38–43.

[3] D. He, Reflection on "8.14" mass blackout in America and Canada one year since then, Power Syst. Technol. 28 (21) (2004) 1–5.

[4] Y. Xu, Inspiration from American and Canadian mass blackout in 2003, Electr. Equip. 5 (2) (2004) 67–69.

[5] Y. Zhao, H. Zhao, Evaluation methods of power grid vulnerability, Shandong Electr. Power (5) (2009) 21–27.

[6] Y. Fan, X. Wang, Q. Wang, New progress of security research on power system – study on vulnerability problems, J. Wuhan Univ. 36 (2) (2003) 110–113.

[7] H. Zhou, D. Liu, Z. Wu, Y. Jiang, H. Cheng, Early warning indicators of power system security and its application, Autom. Electr. Power Syst. 31 (20) (2007) 45–47.

[8] H. Song, M. Kezunovic, "Static Analysis of Vulnerability and Security Margin of the Power System," IEEE 2005 PES Transmission & Distribution Conference & Exposition, Dallas, Texas, May. (2006) 147–152.

[9] D. Watts. Security & vulnerability in electric power systems. NAPS 2003, 35th North American Power Symposium, University of Missouri-Rolla, Rolla, Missouri, 2003, 559–566.

[10] X. Yang, C. Cui, Risk management and early warning of power grid security, Guangxi Electr. Power (3) (2009) 35–38.

[11] D. Ding, Reduce physical vulnerability of power system to cope with natural disasters and sabotage, China Electr. Power 42 (6) (2009) 26–31.

[12] X. Pan, J. Zhang, Vulnerability analysis of power system based on risk assessment, North China Electr. Power Univ. (2008) 20–25.

[13] N. Wang, W. Chen, L. Luo, Early warning of low voltage safety of power system based on risks, East China Electr. Power 36 (3) (2008) 66–69.

[14] W. Chen, Q. Jiang, Y. Cao, Z. Han, Vulnerability assessment of complex power system based on risk theory, Power Syst. Technol. 29 (4) (2005) 12–17.

[15] Q. Li, L. Zhou, Study on Early Warning Models of Voltage Stability of Power System Based on Artificial Immune, Chongqing University, Chongqing, (2007).

[16] G. Zhang, J. Zhang, Q. Peng, The index system and methods of security assessment of power grid, Power Syst. Technol. 33 (8) (2009) 30–34.

[17] H. Xiao, J. Yao, J. Zhang, Analysis on safety assessment system of power grid, Power Syst. Technol. 33 (12) (2009) 77–82.

[18] J. Bai, T. Liu, G. Cao, Review on vulnerability assessment methods of power system, Power Syst. Technol. 32 (2) (2008) 26–30.

[19] X. Wu, J. Liu, P. Bi, Study on voltage stability of distribution network, Power Syst. Technol. 30 (24) (2006) 31–35.

[20] W. Zhou, S. Miao, J. Qu, Y. Zhang, Detailed early warning study on natural disasters of power grid, J. Yunnan Univ. 30 (s2) (2008) 286–290.

[21] Q. Xie, J. Li, The current state and countermeasures of natural disasters in power system, J. Nat. Disasters 15 (4) (2006) 126–131.

CHAPTER 4

Derivation of Distribution Network Vulnerability Indicators Based on Voltage Stability

Contents

4.1 INTRODUCTION

With the continuous development of agriculture and industry, electronic industry and energy, as well as with the changes of the ecological environment, great transformations have taken place in the power grid; for example, the original small and scattered power plants have been gradually replaced by concentrated power plants with large unit capacity, supported by nuclear power, hydropower, and thermal power. Furthermore, most of these power plants are far away from the load center, so the national network power transmission of remote

X. Meng and Z. Pian: Intelligent Coordinated Control of Complex Uncertain Systems for Power Distribution Network Reliability. http://dx.doi.org/10.1016/B978-0-12-849896-5.00004-0

distance and ultra-high voltage is widely used, and the greatest dynamic, nonlinear, and complex artificial system have then appeared – the power grid. The stability of the power grid can be divided into angle stability, frequency stability, and voltage stability [1]. Studies on angle stability have already been very mature, and accidents of frequency stability are few; but, in recent years, voltage stability accidents have occurred frequently [2], attracting more and more people's attention. Voltage instability may cause the whole system to become instable; and, with the rapid development of society, the quantity of load equipment has been substantially increased, calling for higher requirements for the electric power. As such, the study of voltage stability problems caused by load increase is of great significance.

Electricity workers have previously attached great importance to the angle stability issue of power system stability; this reflects the angle changes stability of the generator rotor of relative motion. So, theories and practical application research on angle stability have become relatively mature, and accidents caused by angle instability are now very few, as preventive measures have been organized well. Most of the accidents that occurred in recent years were due to voltage instability.

For example, in 2003, the blackout accident in the joint power grid of Canada and the Northeast and Midwest of the United States of America that lasted about a week, in some places, with the cumulative loss of load reaching 62,000 MW that had caused huge economic losses to the two countries, as well as great inconvenience to the residents' lives [3,4]. Traditional voltage collapses are often caused by a serious shortage of reactive power. The analysis of this accident has found that, before the occurrence of this blackout, the lines were seriously overloaded and led to failure of the control system, and incoordination of the dispatching system, making circuits disconnect one by one, so the chain reactions of power flow transferred, the system oscillated, the voltage at heavy load points collapsed, and the system broke down finally, leading to the serious blackout.

In 2005, a voltage collapse happened in the Moscow power grid [5], and the accident investigation report shows that the aging equipment was the direct reason that led to unit shutdown, important bus disconnection, and overload. The sag of overloaded overhead line increased, flashover discharging on trees and other obstructions, which caused protection actions of the lines. Under such circumstances, the operational personnel didn't take any power rationing measures, therefore, the load of the running line increased and the voltage dropped, leading to a chain development of the accident, and finally the blackout in a large area of the power grid.

In 2006, several western European countries experienced blackout accidents affecting about 10 million people. This was the worst blackout accident of the past 30 years in France. In order to take a large ship out of a factory, Germany cut off two high-tension lines, resulting in an output overload in the eastern European power grid. Meanwhile, the input power of the western power grid was in serious shortage, leading to chain reactions of voltage collapses and, finally, to the big blackout accident occurring.

Voltage instability accidents have also happened in China. In 1989, due to the fast load growth in Guizhou, a blackout accident covering large areas happened, causing the northern Guizhou power grid and the Qingzhen, Guiyang, Shuicheng power plants to step out from the main network, with the loss of load reaching about 480,000 kW. In 1995, a large area of blackout accident happened in Ningxia, with the loss of load reaching about 420,000 kW.

Since the 1980s, the power grid has operated closer and closer to the limit state, significantly increasing voltage instability accidents. The main reasons are: (1) pressure from environmental protection when constructing the power generator and expanding lines; (2) increasing electricity consumption in heavy load areas; (3) new ways of system load in the power market, etc.

Both developed and developing countries have a contradiction among loads, lines, and power generators. Users' loads are increasing constantly, yet the expansion of the power grid is facing greater problems. Due to the heavy load of the power grid, the phenomenon of slow or fast voltage drop happens, and it may lead, in some serious cases, to voltage collapses.

The rapid growth of load, the difficulties in power system expansion, and the occurrence of important blackout accidents and voltage collapse accidents encourage us to probe the cause, characteristics, and mechanism of voltage stability, and to study all kinds of voltage stability indicators, in order to find an indicator that can evaluate the current state of voltage more precisely. For example, whether the voltages of the whole power grid or each node are stable, and how far the whole power grid and each node is from the voltage collapse, etc. Using indicators to guide the practical operation, and putting forward good measures to prevent voltage instability, the scheduling personnel can take advance measures in order to prevent voltage instability and avoid the occurrence of blackout accidents. Hence, studying voltage stability is of great importance.

4.2 OVERVIEW OF VOLTAGE STABILITY

4.2.1 Definition of Voltage Stability

Although there are many researches on defining voltage stability, the essence of these researches is to study the ability of power grid to maintain load voltage level within the specified operating limits [6]. Two kinds of voltage stability definitions are presented subsequently.

In Chinese Standard DL 755-2001 "Guidelines for the Security and Stability of Power Systems," voltage stability is defined as: after a power system suffers various disturbances, including smaller disturbances, such as load rising or decreasing, and larger disturbances such as short circuit, line outage and generating unit tripping, etc., the system voltage can maintain, or recover to, the acceptable range. It also puts forward the forms of voltage instability, including static small disturbance instability, transient big disturbance instability, dynamic big disturbance instability, and instability of longer distances, etc. Finally, the conditions when voltage instability happens are put forward, such as normal operational conditions of the system (the voltage is basically stable at this time), the abnormal operational condition (the node voltage is obviously reduced at this time), and the condition when the system suffers disturbance, etc.

The research report by the joint working group of the IEEE committee and CIGRE38 committee shows that: voltage stability usually refers to the ability of each node to maintain acceptable voltages always after suffering disturbances in a given initial operating condition.

At this time, the balance between load demand and supply plays a decisive role in voltage stability. Once the system voltage stability is destroyed, the node voltage of the system will either increase or reduce, so that transmission lines may suffer successive tripping, the generator falling out of step, and a cascading blackout will happen, resulting in the loss of power of a large number of loads and, eventually, the voltages will collapse. In other words, voltage collapse refers to the situation when the load voltage of the system is falling below the acceptable limit value due to voltage instability [7]. Most instances of voltage instability and voltage collapse are caused by great disturbances, such as load increasing quickly and dramatically, etc.

Voltage instability refers, in general, to continuous and sharp drops in some nodes' voltages of the system, in several seconds, or several minutes after the normally operating power system suffers certain disturbances. At this moment, users cannot receive power as they did normally and, thus, significant instability occurs in the system.

Voltage collapse is caused by voltage instability at the time when most serious consequences happen, usually the result of big disturbances (such as short–circuit, line outage, and the generating unit tripping of electric elements, etc.). Voltage collapse usually refers to the voltages of load nodes or other nodes that are below the acceptable limit value after the normal operating network suffers from certain disturbances. The scope of voltage collapse is different, and it may be global or localized. The two terms, voltage instability and voltage collapse, can be used interchangeably [8]. During voltage collapses, large amount of loads may be lost, which may even make the system split.

4.2.2 Classification of Voltage Stability

Voltage stability can be classified according to different standards, out of which some are based on the nature of the disturbances, and some are based on the scope of research time, as shown in Figure 4.1.

Small disturbance voltage stability refers to a system's ability to keep the voltage within an acceptable range when experiencing disturbances, such as load rising by a small amount. Both load characteristics and influences of continuous and discontinuous control can affect the small disturbance voltage stability. Large disturbance voltage

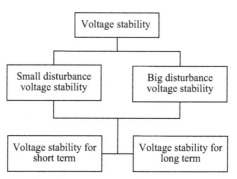

Figure 4.1 Classification of voltage stability.

stability refers to the ability to keep the bus voltage within an acceptable range when experiencing disturbances such as losing the generator, short circuit, line outage, or system failure. System characteristics, load characteristics, as well as the control and protection of the two combined, can determine large disturbance voltage stability. Voltage stability can be divided into long-term voltage stability and short-term voltage stability, according to their period and time length. In general, long-term voltage stability lasts longer, from several minutes to dozens of minutes, a situation that usually studies the generator excitation current limiter and transformer tap adjustment, and so on. The duration of short-term voltage stability is shorter, generally lasting a few seconds, mainly studying the HVDC converter and induction motor [9].

4.2.3 Introduction of Voltage Stability Terms

At the present time, unstable power grid accidents happened frequently; out of these, accidents caused by voltage instability take up a great part, so research on voltage stability is imperative. In recent years, research on problems concerning voltage stability is not yet mature. The summarization of some voltage stability terms follows [10,11].

Voltage stability: refers to the ability of the system voltage to keep or restore to the limit value, depending on its own characteristics, such as modes of operation, line parameters, reactive power compensation device, onload tap changer, transformer tap, and other devices' control effects at the time when the power system suffers small or large disturbances. As the load admittance received by the system increases, the load power also increases, and the whole system power and voltage can be controlled.

Voltage collapse: when the system experiences certain disturbances, the equilibrium state of the network reactive power will no longer exist, and adopting adjustment and control measures meanwhile cannot make the voltage near the lode node maintain or recover to the allowed range, resulting in an irreversible declining process of local or global power grid voltage.

Voltage instability: when the current system does not meet the conditions of the voltage stability, the node voltage at this time may be rising or decreasing continuously, resulting in the transmission power of the system exceeding the maximum transmission limit, and the load power going back to oscillation, a situation that will further reduce the power consumption. Voltage instability may occur when the system voltage is either normal or abnormal, and it is more likely to occur after suffering disturbances in the system.

Voltage stability limit: when all the load values of the system reach a certain level, if an additional load is added, the load node voltage might fall dramatically, leading to voltage collapse; at this moment, the total value of the load power endured by the system is considered the maximum value that the system can bear, also known as the voltage stability limit. This is also an indicator used to measure whether the voltage is stable or not.

4.2.4 Influential Factors of Voltage Stability

Inherent factors affecting voltage stability are:
1. The load characteristics. Load characteristics are the most direct and critical influential factors that are relatively active, such as the increasing speed of load, load type, static and dynamic characteristics of load, and so on. If the load is growing so fast that it exceeds the maximum power the system can bear, voltage may thus lose stability. For example, in loads, the influence of main asynchronous motor on voltage stability is great, both in heavy loads and light loads. If the system voltage drops, the reactive power absorbed by an asynchronous motor will also

decline accordingly, yet more seriously as the voltage declines further; the reactive power absorbed by an asynchronous motor may increase, making the voltage instability even worse. Since the influence of the load static characteristic ZIP model on voltage stability is different, the factors considered by dynamic characteristics are greater.

2. The ability of transmission power of the power grid, including the structure of the power grid, parameters and operation modes, etc.

3. Devices related to voltage stability, such as control devices, reactive power compensation devices, the reactive power generator, protection equipment, transformer tap adjustment and onload tap changer, etc., whose timely and effective interactions have a great influence on voltage stability.

There are other factors impacting on voltage stability, and the following factors can still lead to voltage stability problems, and make them even worse:

1. The distance between the load center and the energy base is too long, leading to the long distance transmission of electric power, and the transmission power of the transmission lines, as well as the increasing reactance, will cause part of the lines to overloaded, gradually leading the line to trip, and even to result in voltage collapse, in some serious cases.

2. In recent years, small power plants have gradually been replaced by large ones, with larger unit capacity, making its synchronous reactance increase, and its inertial time constant of the unit decrease. The rise in generator synchronous reactance will make the transmission lines' power limit decrease, which will also cause stability problems.

3. Due to the long distance between load and power generator, as well as the increasing loads, the capacity of transmission lines increases. In this case, if lines are broken because of accidents, a large power vacancy will occur at the sending and receiving ends, seriously threatening the power stability.

4. The increase in transmission line circuit makes the possibility of multiple failures of lines rise.

In domestic usage, voltage instability and voltage collapse problems also exist, though they are not serious; and voltage stability problems are always hidden, as the taps of many onload tap changers haven't started to switch automatically, and scheduling workers often take load shedding measures in case of an emergency – this will lead to blackout, bringing inconvenience to users' lives and to production. So, this measure should not be used more often or at the earlier stage, otherwise it will reduce the reliability of the power supply.

4.3 EVALUATION INDICATOR *L* OF VOLTAGE STABILITY IN THE POWER GRID

In the current–voltage stability study, when the voltage is so low that the system cannot bear it, and other adjusting measures can't be effective, scheduling workers usually take load-shedding measures in order to avoid voltage collapse accidents; this will directly reduce the power supply reliability of the system. Studying voltage stability is necessary in order to find out the indicator that is able to judge whether the voltage is stable or not. This indicator is used to measure the stability of the power grid voltage, so, before defining the indicator, it is necessary to study the nature of voltage collapse thoroughly. The key factor leading to voltage collapse is that the equilibrium point between the supply and demand of the system reactive power does not exist. Voltage collapse occurs commonly in heavy load systems when the unbalance between power generation and transmission plans of the system appears, leading to a continuously growth in load before the collapse. Besides, small disturbances of system voltage, such as load rising and breaking lines, will also lead to voltage collapse, and the node voltage of the system decreases continuously until voltage collapse finally appears.

Therefore, it is of great importance to find an accurate and reliable indicator to reflect whether the voltage is stable or not, and, based on it, scheduling workers can judge whether the voltage of the system or of each node is stable, which are the weak points of voltage, and that how far the current state is from voltage collapse, and so on. This chapter adopts the indicator L to evaluate load voltage stability, and the derivation process is illustrated in detail subsequently.

4.3.1 Overview of Voltage Stability Assessment Indicator *L*

1. Definition of voltage stability assessment indicator *L*

The voltage stability assessment indicator is used to evaluate whether the voltage of a load node in the system is stable or not, so it is called indicator *L* [12]. This indicator is achieved based on power flow calculation, and it belongs to the state indexes. Since only the load node voltage is evaluated in the system, it also falls into local indicators. The calculating process is also made through extension from the simple system to the complex system.

2. Characteristics of voltage stability assessment indicator *L*

Voltage stability assessment indicator *L* is derived from a simple two-node system, at first; this is a system containing one generator and one load node, and it is then extended to a complex system. In a word, during the solving process, the indicator divides all the nodes in the system into two types, generator node (*PV* node) and load node (*PQ* node), and after the indicator's expression is deduced, the value of the indicator can be achieved by using power flow calculation results.

Indicator value *L* generally falls in the range of [0, 1], and, according to the size of distance between *L* and 1, the voltage stability degree can be seen visually, that is, when *L* < 1, the system is stable. The closer *L* is to 1, the more unstable the node voltage is. But when *L* = 1, voltage collapse will occur. For each load node in the system, this indicator value gives its degree of voltage stability, respectively, and the stability of the whole system can be measured by using the maximum indicator value of all nodes in the system. As such, this indicator is intuitively clear, and the calculation is simple and fast.

The reason why this section uses this indicator is that it is simple, has a fast calculation speed, and it is suitable for online applications. In recent years, loads have increased significantly, and voltage instability accidents due to increasing load are becoming more common. This indicator is also a local indicator that is able to measure an element or a node. Besides, a load node is an important link of the power system, reflecting directly the reliability of power supply, so this indicator can calculate the voltage instability caused by the increasing load, very accurately and quickly, a fact that is widely used in the localized regional load growth of the complex system.

4.3.2 Derivation of Voltage Indicator *L* in Multinode System

1. Derivation of voltage indicator *L* in two–node system

The equivalent circuit of a two-node system is shown in Figure 4.2. From the figure, it can be seen

$$Y_s \dot{V}_2 + Y_L \left(\dot{V}_2 - \dot{V}_1 \right) = \dot{I}_2 = \frac{S_2^*}{\dot{V}_2^*} \tag{4.1}$$

$$S_2 = \dot{V}_2 \dot{I}_2^* \tag{4.2}$$

From expression (4.1), and expression (4.2), it can be derived

$$Y_s \dot{V}_2 \dot{V}_2^* + Y_L \dot{V}_2 \dot{V}_2^* - Y_L \dot{V}_1 \dot{V}_2^* = S_2^* \tag{4.3}$$

$$V_2^2 \left(Y_S + Y_L \right) - Y_L \dot{V}_1 \dot{V}_2^* = S_2^* \tag{4.4}$$

For

$$Y_{22} = Y_L + Y_S, Y_{21} = -Y_L$$

So,

$$V_2^2 Y_{22} + Y_{21} \dot{V}_1 \dot{V}_2^* = S_2^*$$
$$V_2^2 + \frac{Y_{21}}{Y_{22}} \dot{V}_1 \dot{V}_2^* = \frac{S_2^*}{Y_{22}} \tag{4.5}$$

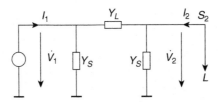

Figure 4.2 Equivalent circuit diagram of the node system.

Make $\dfrac{Y_{21}}{Y_{22}}\dot{V}_1 = \dot{V}_0$, so

$$V_2^2 + \dot{V}_0\dot{V}_2^* = \frac{S_2^*}{Y_{22}} \tag{4.6}$$

Make $\dot{V}_0 = V_0 \angle \delta_0, \dot{V}_2 = V_2 \angle \delta_2$, so

$$V_2^2 + \dot{V}_0\dot{V}_2^* = V_2^2 + V_0V_2 e^{j(\delta_0 - \delta_2)} = \frac{S_2^*}{Y_{22}} = a + jb \tag{4.7}$$

For

$$
\begin{aligned}
V_2^2 + V_2V_0 e^{j(\delta_0 - \delta_2)} &= V_2^2 + V_2V_0 \cos\left(\delta_0 - \delta_2\right) \\
&+ jV_2V_0 \sin\left(\delta_0 - \delta_2\right) = a + jb
\end{aligned}
\tag{4.8}
$$

Make $\theta = \delta_0 - \delta_2$, so

$$
\left.
\begin{aligned}
f_1(V_2, \theta) &= V_2^2 + V_0V_2 \cos\theta = a \\
f_2(V_2, \theta) &= V_0V_2 \sin\theta = b
\end{aligned}
\right\}
\tag{4.9}
$$

In expression (4.9), differentiate V_2 and θ respectively and it can be seen

$$
\begin{cases}
\dfrac{\partial f_1}{\partial V_2} = 2V_2 + V_0 \cos\theta & \dfrac{\partial f_2}{\partial V_2} = V_0 \sin\theta \\[2mm]
\dfrac{\partial f_1}{\partial \theta} = -V_0V_2 \sin\theta & \dfrac{\partial f_2}{\partial \theta} = V_0V_2 \cos\theta
\end{cases}
\tag{4.10}
$$

So, the Jacobian matrix is

$$
J = \left\{
\begin{matrix}
2V_2 + V_0 \cos\theta & -V_0V_2 \cos\theta \\
V_0 \sin\theta & V_0V_2 \cos\theta
\end{matrix}
\right\}
\tag{4.11}
$$

When the voltage collapses upon a neighboring point, the determinant of the Jacobian matrix is zero, so

$$\det(J) = \left(2V_2 + V_0 \cos\theta\right)V_0V_2 \cos\theta - \left(-V_0V_2 \sin\theta\right)V_0 \sin\theta = 0 \tag{4.12}$$

Therefore,

$$2V_2 \cos\theta = V_0 \rightarrow \frac{V_2}{V_0}\cos\theta = -\frac{1}{2} \tag{4.13}$$

It is able to deduce

$$\text{Re}\left\{\frac{\dot{V}_2}{\dot{V}_0}\right\} = -\frac{1}{2} \tag{4.14}$$

Divide the two sides of expression (4.6) by V_2^2 at the same time

$$\left|1+\frac{\dot{V}_0}{\dot{V}_2}\right| = \frac{S_2^{\star}}{Y_{22} \times V_2^2} \tag{4.15}$$

From expression (4.14) and expression (4.15), it can be derived

$$\left|1+\frac{\dot{V}_0}{\dot{V}_2}\right| = |1+(-2)| = 1 \tag{4.16}$$

The value of the indicator is 1 when the voltages collapse.

From this, the indicator L of load node 2 can be derived when it is close to the voltage collapse:

$$L_2 = \left|1+\frac{\dot{V}_0}{\dot{V}_2}\right| = \left|\frac{S_2^{\star}}{Y_{22} \times V_2^2}\right| \tag{4.17}$$

So, $0 \le L \le 1$

2. Voltage indicator L's derivation in multinode system

Based on a two–node system, the derivation process of voltage indicator L in the complex system is as follows:

Because $I = YU$ and the system nodes are divided into PQ node and PV node, therefore,

$$\begin{bmatrix} I_L \\ I_G \end{bmatrix} = \begin{bmatrix} Y_{LL} & Y_{LG} \\ Y_{GL} & Y_{GG} \end{bmatrix} \begin{bmatrix} V_L \\ V_G \end{bmatrix} \tag{4.18}$$

It can be deduced

$$
\begin{bmatrix} V_L \\ I_G \end{bmatrix} = \begin{bmatrix} Z_{LL} & -Z_{LL}Y_{LG} \\ Y_{GL}Z_{LL} & Y_{GG} - Y_{GL}Z_{LL}Y_{LG} \end{bmatrix} \begin{bmatrix} I_L \\ V_G \end{bmatrix} \tag{4.19}
$$

In the expression, V_L, I_L represent the voltage and current vectors of node PQ; V_G, I_G represent the voltage and current vectors of node PV.

From expression (4.19), it can be seen that for any load node i, it has

$$
\dot{V}_i = \sum_{j \in A_L} Z_{ij} \times \dot{I}_j + \sum_{k \in A_G} M_{ik}\dot{V}_k \left(i \in A_L, k \in A_G \right) \tag{4.20}
$$

In the expression $M_{ik} = -Z_{ij}Y_{ik}$; A_L is the number of the PQ node; A_G is the number of the PV node; $\dot{V}$ is the voltage of load node i; $\dot{V}_k$ is the voltage of the PV node; $\dot{I}_j$ is the current of the load node.

Assume $\dot{V}_0 = -\sum_{k \in A_G} M_{ik}\dot{V}_k$, and multiply the two sides of expression (4.20) by $\dot{V}_i^*$

$$
V_i^2 = \dot{V}_i^* \sum_{j \in A_L} Z_{ij} \times \dot{I}_j - \dot{V}_0 \dot{V}_i^* \tag{4.21}
$$

Make $\dot{V}_i^* \sum_{j \in A_L} Z_{ij} \times \dot{I}_j = \dfrac{S_i^{+*}}{Y_{ii}^+}$, of which $S_i^+ = S_i + S_i^e, S_i^e = \sum_{\substack{j \in A_L \\ j \neq 1}} \dfrac{V_i^* \dot{I}_j}{Y_{ij}}$

$$
V_i^2 + \dot{V}_0 \dot{V}_i^* = \frac{S_i^{+*}}{Y_{ii}^+} \rightarrow 1 + \frac{\dot{V}_0}{\dot{V}_i} = \frac{S_i^{+*}}{V_i^2 Y_{ii}^+} \tag{4.22}
$$

S_i^+ is the sum of the load power of node i itself and other load node power connected to node i.

It can be seen that expression (4.15) and expression (4.22) have the same format, so the voltage stability indicator L of load node i in the complex system is

$$
L_i = \left| 1 + \frac{\dot{V}_0}{\dot{V}_i} \right| = \left| \frac{S_i^{+*}}{V_i^2 Y_{ii}^+} \right| \tag{4.23}
$$

Above is the derivation process of the voltage stability indicator L, based on a power flow solution, and the following chapters will use this indicator to establish a voltage stability assessment model of the power grid, based on the neural network and particle swarm optimization neural network algorithm; thus, it will be possible to evaluate quickly whether the grid voltage is stable or not.

4.4 THE MODEL FOR INDICATOR CALCULATION

Although the indicator L has been deduced to evaluate the voltage stability of the load node in the power grid, a fast calculating model is still needed, as the traditional calculating method takes too much time when the nodes are large or changing fast in the power grid. As such, this chapter introduces particle swarm optimization to improve the BP network, and establishes models to solve the voltage stability indicator L when loads are changing very fast.

4.4.1 Combination Point of the PSO and the BP network

1. The relationship of the position vector of the particle swarm, with weights and threshold values of the neural network. When the BP network is training, it is connected with weights and threshold values to search for a better value by using particle swarm optimization; this will improve the training speed, as well as the precision. Each particle in the particle swarm is expressed by a position vector and a speed vector, with position vectors mapping the weights and threshold values of the neural network, whose detailed combination method is as follows:

    Adopt a three-layer BP network, and n, m, R is the number of neurons of the input layer, the output layer, and the hidden layer, respectively, so the number of threshold values of the hidden layer and output layer is n, m, and the digit capacity N of the personal particle position vector is the sum of the number of BP network weights and threshold values that is

$$N = n \times m + m \times R + n + m \tag{4.24}$$

2. The speed vector of the particle swarm. In general, the particle speed falls into the defined range of $[v_{max}, v_{min}]$. If the real speed during the particle's flying process exceeds the maximum speed, its speed will be limited to v_{max}, and if it is slower than the minimum speed, it will be limited to v_{min}.

3. The fitness functions of the particle swarm. Because the initial weights and threshold values of the BP network are optimized, the fitness function *fitness* of the particle swarm is the function of error sum of squares of the BP network, whose expression is:

$$fitness = \frac{1}{2n_s} \sum_{j=1}^{n_s} \sum_{k=1}^{m} \left(Y_{jk} - y_{jk} \right)^2 \tag{4.25}$$

In this expression, n_s is the input and output logarithm of the samples; m is the number of neurons at the output end.

In conclusion, the combination process of a PSO and BP network is: the position of each particle in the particle swarm is expressed by the vector composed by all the weights and threshold values of the BP network, with the vector number of weight and threshold values common to the BP network equivalent to the vector dimensions of individual position of particles; this means that each individual's weight in the particle swarm is mapping the weights and threshold values of the BP network, forming the whole BP network, and training by inputting the training samples. The optimization process of weights and threshold values is an iterative process.

4.4.2 Combination Procedure of the PSO and BP Network and Its Parameter Setting

The updating formula of speed and position of the particle are formulae (2.14) and (2.15), and the combination procedure of the PSO and BP network is shown in Figure 4.3.

Assume the number of particle swarm is 40, whose maximum speed $v_{max} = 0.5$, so $v_{min} = -0.5$, and if the speed exceeds

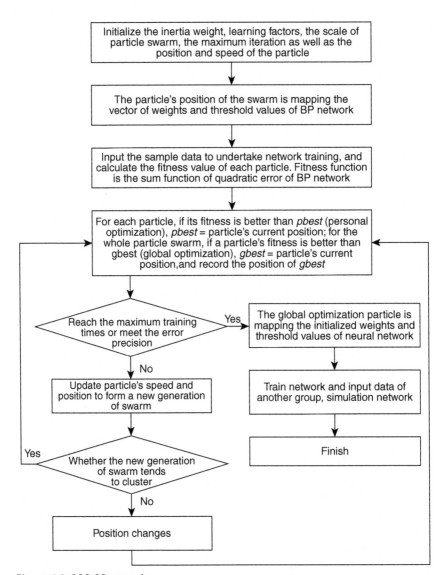

Figure 4.3 PSO-BP procedure.

the extremum, then it should be limited as the corresponding maximum speed or minimum speed. The maximum iteration $v_{min} = -0.5$, the learning factor $c_1 = c_2 = 2$, the inertia weight $w_{max} = 0.9, w_{min} = 0.4$, from the experiment, it can be seen that the results adopted first increased, and then decreased strategy on the input sample data is better, and at this moment the expression of inertia weight is:

$$w(t) = \begin{cases} 1 \times \dfrac{t}{t_{max}} + 0.4 \left(0 \le \dfrac{t}{t_{max}} \le 0.5 \right) \\[4mm] -1 \times \dfrac{t}{t_{max}} + 1.4 \left(0.5 \le \dfrac{t}{t_{max}} \le 1 \right) \end{cases} \tag{4.26}$$

Of which t is a value ranging from $1 \sim t_{max}$. Adopt the strategy of first increased and then decreased, so the speed of early convergence is faster, and it is able to search a larger space, with a strong local searching ability on the later stage. It is a better improvement method because it reaches global and local equilibrium by keeping the merits of decreasing and increasing strategies to a large extent.

4.5 EXAMPLE ANALYSIS

4.5.1 Five-Node Circuit Diagram and Known Parameters

First, the five-node system needs to be adopted to train and forecast. The artificial circuit of five-node system is shown in Figure 4.4, of which 3, 4, 5 of the system are three load nodes, node 1 and 2 are generator node and swing bus, and the line parameters are shown in Tables 4.1 and 4.2.

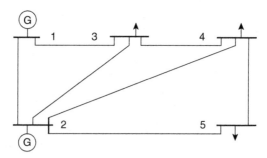

Figure 4.4 Simulation circuit of the five-node system.

Table 4.1 Parameters of five-node system lines

Branch number	Lead node	End node	Branch resistance	Branch reactance	1/2 ground susceptance
1	1	2	0.02	0.06	0
2	1	3	0.08	0.24	0
3	2	3	0.06	0.18	0
4	2	4	0.06	0.18	0
5	2	5	0.04	0.12	0
6	3	4	0.01	0.03	0
7	4	5	0.08	0.24	0

Table 4.2 Nodes Data of a five-node system

Node number	Node category	Node voltage	Generator output		Load power	
			Active	Reactive	Active	Reactive
1	Swing bus	1.0600	–	–	0.0000	0.0000
2	PV node	1.0000	0.20000	0.20000	0.0000	0.0000
3	PQ node	1.0000	0	0	0.4500	0.1500
4	PQ node	1.0000	0	0	0.4000	0.0500
5	PQ node	1.0000	0	0	0.6000	0.1000

4.5.2 BP Algorithm of Five-Node Samples and Improved BP Algorithm Simulation

1. Input data. The input data includes input samples and desired output samples, and it can be seen from Figure 4.4 that the five-node system's input contains the active and reactive power, the PV node power, and voltage of three load nodes: 3, 4, and 5, as well as the voltage of swing bus, with ten in total. The output contains the voltage of the PQ node and the voltage index L, with six in total. Constantly increase the loads to form an input sample, and 40 groups of samples are chosen here, yet there will be a list of only 20 groups due to space constraints. Part of the training sample data are in Table 4.3. Adopting the results of the power flow calculation, desired outputs write the calculation

Table 4.3 Part of the training sample data of a five-node system

Load	Input samples						Desired output samples			
F	PL3	QL3	PL4	QL4	V1	V2	V3	V4	L3	L4
1	0.45	0.15	0.4	0.05	1.06	1	1.0088	1.0073	0.0564	0.0432
1.1	0.495	0.165	0.44	0.055	1.06	1	1.0016	0.9998	0.0631	0.0511
1.2	0.54	0.18	0.48	0.06	1.06	1	0.9942	0.9922	0.0689	0.0612
1.3	0.585	0.195	0.52	0.065	1.06	1	0.9865	0.9842	0.0732	0.0712
1.4	0.63	0.21	0.56	0.07	1.06	1	0.9785	0.976	0.0823	0.081
1.5	0.675	0.225	0.6	0.075	1.06	1	0.9703	0.9674	0.09	0.085
1.6	0.72	0.24	0.64	0.08	1.06	1	0.9616	0.9585	0.0955	0.0951
1.7	0.765	0.255	0.68	0.085	1.06	1	0.9526	0.9492	0.1086	0.1037
1.8	0.81	0.27	0.72	0.09	1.06	1	0.9432	0.9395	0.1679	0.1123
1.9	0.855	0.285	0.76	0.095	1.06	1	0.9334	0.9293	0.2156	0.1598
2	0.9	0.3	0.8	0.1	1.06	1	0.923	0.9185	0.2435	0.1745
2.1	0.945	0.315	0.84	0.105	1.06	1	0.912	0.9072	0.2823	0.1823
2.2	0.99	0.337	0.88	0.11	1.06	1	0.8994	0.8939	0.2932	0.2013
2.3	1.035	0.345	0.92	0.115	1.06	1	0.888	0.8823	0.2993	0.252
2.4	1.08	0.36	0.96	0.12	1.06	1	0.8747	0.8685	0.3075	0.2723
2.5	1.125	0.375	1	0.125	1.06	1	0.8603	0.8536	0.3189	0.2885
2.6	1.17	0.39	1.04	0.13	1.06	1	0.8445	0.8373	0.3325	0.3012
2.7	1.215	0.405	1.08	0.135	1.06	1	0.8269	0.819	0.3521	0.3225
2.8	1.26	0.42	1.12	0.14	1.06	1	0.8069	0.7982	0.3967	0.4245
2.9	1.305	0.435	1.16	0.145	1.06	1	0.783	0.7735	0.4321	0.4776

program of voltage indicator L. If the system is too large, PSD-BPA or PSASP can be adopted to calculate the power flow, and then an index calculation program is used to calculate the values of indicators.

2. Determine the number of neurons in each layer of the BP network. From the above, the dimension of the input sample of a five-node system is 10, and the dimensions of the desired output sample is 6, so the number of neurons in input layer and output layer of the BP network is 10 and 6, respectively, since the number of neurons in the input layer and output layer is determined by the dimension of input quantity and desired output quantity. The number of neurons in the hidden layer is achieved through constant experiments, and their number is 12 in the five-node system.

So, the model of a three-layer BP neural network of five-node system is 10-12-6.

3. Determine other parameters in the BP network. Since the improved BP network is adopted that uses a variable learning rate method, the initialized learning rate is then set 0.04, changing constantly during the training process. Besides, momentum factors are adopted, and the initialized momentum factor is 0.8. The training should reach the precision of 0.00001, with different training times.

4. Training results of the BP network and the PSO optimized BP network of the five-node system. After the setup of the initialized parameters, the BP network training process will be prepared to train, and the normalization process of sample data will be under way during the training process. The training curve of the BP network in five-node system is shown in Figure 4.5. The speed of the BP network training is slow with such a low precision, and so it doesn't meet the requirements. Due to the randomness employed in choosing initialized weights and threshold values during the BP network training process, the training speed of the network is slow, and with low precision. So, a particle swarm optimized BP network improves the training speed and precision, using the same 40 groups of input data sample with the BP network. The PSO optimized curve of a five-node system is shown in Figure 4.6, and the training curve after the BP network is optimized is shown in Figure 4.7.

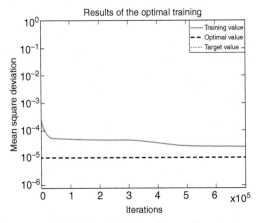

Figure 4.5 Training curve of the BP network in a five-node system.

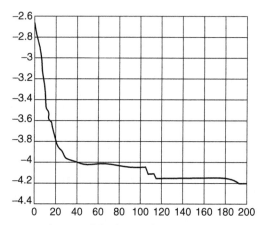

Figure 4.6 PSO optimized curve of the five-node system.

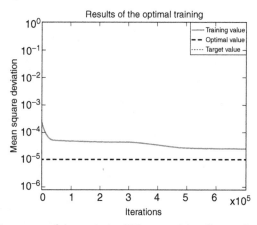

Figure 4.7 Training curve of the optimized BP network in a five-node system.

Since the sample data are too large to be presented here in full, the training result data of part of the nodes are listed. In a five-node system, due to the large initialized load values of lode node 3 and 5, the possibility of voltage instability is greater if the load is increased, so here the training results of the BP network and the PSO optimized BP network of node 3 and 5 are listed, and the comparison results are in Table 4.4 and Table 4.5.

5. Forecasting results of the BP network and the PSO optimized BP network in five-node system. After the BP network and the PSO optimized BP network are trained, their generalized ability,

Table 4.4 Comparison table between training results and expected results of a BP network and an optimized BP network of load node 3

Load	BP voltage	Expected voltage	Optimized BP voltage	BP index	Expected index	Optimized BP index
1	1.0038	1.0088	1.0088	0.0544	0.0564	0.0564
1.1	1	1.0016	1.0016	0.053	0.0631	0.0631
1.2	0.98	0.9942	0.9942	0.0679	0.0689	0.0689
1.3	0.976	0.9865	0.9865	0.0735	0.0732	0.0732
1.4	0.9781	0.9785	0.9788	0.0821	0.0823	0.0824
1.5	0.9688	0.9703	0.97031	0.08	0.09	0.094
1.6	0.9609	0.9616	0.9616	0.0946	0.0955	0.0955
1.7	0.951	0.9526	0.9526	0.108	0.1086	0.1086
1.8	0.94	0.9432	0.9432	0.01678	0.01679	0.01679
1.9	0.9311	0.9334	0.9334	0.2158	0.2156	0.2157
2	0.915	0.923	0.923	0.2434	0.2435	0.2435
2.1	0.901	0.912	0.912	0.2814	0.2823	0.2823
2.2	0.8982	0.8994	0.8994	0.2911	0.2932	0.2932
2.3	0.881	0.888	0.888	0.299	0.2993	0.2993
2.4	0.8721	0.8747	0.8747	0.3069	0.3075	0.3075
2.5	0.83	0.8603	0.8603	0.3189	0.3189	0.3189
2.6	0.842	0.8445	0.8445	0.3321	0.3325	0.3325
2.7	0.8254	0.8269	0.8269	0.3513	0.3521	0.3521
2.8	0.8053	0.8069	0.8069	0.3963	0.3967	0.3967
2.9	0.762	0.783	0.783	0.4315	0.4321	0.4321

also known as network forecast ability, needs to be tested. Here, the so called network generalization ability refers to the ability of the network that has been trained by the training samples to check its response on data of different training samples that reflects the influence of the network structure unpredictability, and samples variability on the network forecasting ability. Here, 15 groups of input data, which are different from the training sample, are chosen to make the network stay in a working state. At the moment the testing program should be prepared, and the output data can be obtained by calling the trained networks during the testing procedure. Since the changes of node 5 voltages and of indicator value L are large, the test results only list node 5 data, with output data comparisons of the test in Table 4.5. The expected data in the table are also achieved through program calculation.

Table 4.5 Comparison table between forecasting results and expected results of a BP network and an optimized BP network of load node 5

Load	BP voltage	Expected voltage	Optimized BP voltage	BP index	Expected index	Optimized BP index
1.05	0.9963	0.9974	0.9974	0.0779	0.0786	0.0786
1.27	0.9778	0.9784	0.9784	0.0935	0.0943	0.0947
1.33	0.9726	0.9730	0.9729	0.0968	0.0976	0.0977
1.48	0.9580	0.9588	0.9588	0.1096	0.1092	0.1096
1.53	0.9525	0.9538	0.9538	0.1146	0.1143	0.1146
1.62	0.9422	0.9447	0.9447	0.1280	0.1340	0.1330
1.73	0.9371	0.9416	0.9416	0.2067	0.2091	0.2090
1.88	0.9150	0.9163	0.9163	0.2683	0.2745	0.2744
1.99	0.9011	0.9027	0.9027	0.2981	0.3101	0.3101
2.12	0.8852	0.8859	0.8858	0.3858	0.3973	0.3973
2.43	0.8410	0.8394	0.8393	0.4801	0.4864	0.4860
2.56	0.8153	0.8161	0.8161	0.5414	0.5432	0.5432
2.78	0.7630	0.7639	0.7638	0.6298	0.6328	0.6327
2.98	0.7032	0.7039	0.7039	0.7423	0.7431	0.7430
3.05	0.6676	0.6685	0.6685	0.7845	0.7866	0.7865

Training and forecasting work have all been done, but only part of the training and forecasting data are listed, due to the large volume. From Table 4.5, it can be noted that the precision of the BP network forecasting result is relatively low; it however reaches the requirements after it is optimized by PSO, though the training speed is relatively slow. Due to the small size of the network, the line connection form of a five-node system is relatively simple, so the size of the voltage stability indicator L is affected by its load size. From the initialized circuit parameters, it can be seen that the load of node 5 is the biggest, and by observing the training results and the forecasting results, it can be found that its voltage indicator L value is also bigger. In Table 4.5, when the load is 3.05 times that of the initialized load, the node 5 voltage indicator L is 0.7865, very close to 1, so a very small disturbance (such as an increased load) may lead to voltage collapse of node 5, when it is at the power consumption peak; this will make the whole system lose voltage stability, so node 5 must be included by scheduling personnel in the key monitoring range.

BIBLIOGRAPHY

[1] Y. Su, S. Cheng, J. Wen, et al., Voltage stability and its research status of the power system (I), Power Syst. Autom. Equip. 26 (6) (2006) 97–101.

[2] X. Hu, Reflection and enlightenment of large blackout accidents of the United State and Canada combined power grid, Power Syst. Technol. 27 (9) (2003) T3–T6.

[3] J. Qu, J. Guo, Statistical analysis of power grid accidents in China during the "Ninth Five-Year" period, Power Syst. Technol. 28 (21) (2004) 60–63.

[4] Y. Yu, C. Dong, The voltage collapse during "8.14" blackout accidents in the United States and Canada, Power Syst. Technol. Elec. Equip. 5 (3) (2004) 4–7.

[5] S. Lu, L. Gao, K. Wang, et al., The analysis and enlightenment of Moscow blackout accidents, Relay 34 (16) (2006) 27–31.

[6] Z. Han, The Stability of Power System, China Electric Power Press, Beijing, (1995).

[7] IEEE/CIGRE Joint Task Force, Definition and classification of power system stability, IEEE Trans. Power Syst. 19 (3) (2004) 1378–1401.

[8] CIGRE Task Force 38.02.10, Modeling of voltage collapse including dynamic phenomena, Electra (147) (1993) 71–77.

[9] IEEE Committee Report, Proposed terms and definition for power system stability, IEEE Trans. App. Syst. (101) (1982) 1894–1899.

[10] C.W. Taylor, Power System Voltage Stability, McGraw-Hill, New York, (1994).

[11] H. Jia, X. Yu, Y. Yu, An improved voltage stability index and its application, Int. J. Electr. Power Energy Syst. 27 (8) (2005) 567–574.

[12] P. Kessel, H. Glavitch, Estimating the voltage stability of power systems, IEEE Trans. Power Delivery 1 (3) (1986) 346–354.

CHAPTER 5

Vulnerability Assessment of the Distribution Network Based on Quantum Multiagent

Contents

*X. Meng and Z. Pian: Intelligent Coordinated Control of Complex Uncertain Systems for Power
Distribution Network Reliability.* http://dx.doi.org/10.1016/B978-0-12-849896-5.00005-2
Copyright © 2016 China Electric Power Press. Published by Elsevier Inc. All rights reserved.

5.1 INTRODUCTION

The security evaluation method of a power system can be used in its vulnerability analysis, since the vulnerability assessment of a power system is also the security analysis of that power system, that is, to find the weakness both inside and outside of the power system and to get the degree of antidisturbance, and the ability to maintain normal operation of each element in the system.

The potential vulnerability in power systems can be divided, in general, into external vulnerability and internal vulnerability. External vulnerability includes sabotage, misoperation, and natural disasters, etc., yet internal fragility comes from the power system itself, including failures of the information communication system, as well as the imperfect evaluation and decision-making of the system. When some accidental factors stack together in a specific environment, it will often lead to the breakdown of some weak point in the network, and transient changes (such as the transient load peak of voltage, frequency, etc.) caused by this load will immediately affect other nodes connected with it' meanwhile, these connected nodes may also have their own vulnerabilities, whose joint effects can lead to successive actions of protective relaying equipment and automatic safety devices, making the fault influence spread and expand, forming a cascading collapse that will eventually lead to a system disaster covering a large area [1].

The existing distribution network usually adopts a centralized control system structure that requires the control center to collect data from multiroutes far away. So, when a fault occurs, it will be hard for the control center to implement timely and effective interferences in a short period of time. Meanwhile, in the operating process, the system also needs a preventive strategy to find and isolate the hidden faults. Therefore, an intelligent distributed system is needed to control randomly the complex power system in real-time, and the multiagent system has provided an idea to realize this kind of system [2].

This chapter first studies the multiagent coordinated control theory, and puts forward a brand new multiagent coordinated learning algorithm based on quantum (Q-MAS) in order to overcome the curse of dimensionality of actions and states, as well as the problems of slow learning speed during multiagent collaborative learning, a fact that has improved multiagent learning speed greatly. Then, in view of the vulnerability assessment and control problems of the distribution network, the multiagent coordinated control theory is applied in the vulnerability analysis of the distribution network in order to design a hierarchical control structure based on Q-MAS that realizes the coordination of vulnerability assessment of the whole distribution network.

5.2 REINFORCEMENT LEARNING OF THE MULTIAGENT SYSTEM BASED ON QUANTUM THEORY

Having been applied in the computer intelligence, quantum algorithms were first used in the simple expert system, applied since in many areas, such as quantum associative memory, artificial neural network and fuzzy logic, and so on.

5.2.1 Some Concepts Used in the Reinforcement Learning of Quantum Multiagent System [3–5]

1. Qubit

   Qubit is the basic unit of quantum information, which is a random superposed state of two-state quantum system, marked as $|\psi\rangle = \alpha|0\rangle + \beta|1\rangle$ that satisfies $|\alpha|^2 + |\beta|^2 = 1$. $|0\rangle$ and $|1\rangle$ correspond to the logic state 0 and 1. $|\alpha|^2$ and $|\beta|^2$ represent the occurrence probabilities of logic state 0 and 1.

2. Hadamard gate

   Hadamard gate is also known as H gate, which is one of the most frequently used quantum gates, recorded as $H \equiv \dfrac{1}{\sqrt{2}}\begin{bmatrix} 1 & 1 \\ 1 & -1 \end{bmatrix}$. Hadamard gate can be used to convert the qubit from clustering state to uniform superposed state.

3. Grover operator

   Grover operator is an important operation of quantum search algorithm put forward by Grover that can increase the amplitude of the target vector, and decrease the amplitude of nontarget vector through iterations.

5.2.2 Algorithm Analysis

N_s and N_a represent the state and action inside the agent, respectively, and figures m and n are selected, and they fall into the range of $N_s \leq 2^m \leq 2N_s, N_a \leq 2^n \leq 2N_a$, finally, m and n qubits are used to represent the state set S and action set A. So

$$S: \left[\begin{array}{c|c|c|c} a_1 & a_2 & & a_m \\ \hline b_1 & b_2 & \cdots & b_m \end{array} \right] \left[\text{of which } |a_i|^2 + |b_i|^2 = 1; i = 1, 2, \cdots, m \right] \quad (5.1)$$

It can also be expressed as $S: \left| \overbrace{00\cdots0}^{m} \right\rangle$.

S is the internal world mode of the agent, and the agent selects actions according to the verification data achieved from S and the external world. Each state s in the state set S is expressed by the superposed state of corresponding number of qubits, with each vector standing for the probability of the corresponding action. When one of these vectors' probability amplitude increases to a certain degree, the learning process is finished.

Set A is expressed as:

$$A: \left[\begin{array}{c|c|c|c} a_1 & a_2 & & a_n \\ \hline b_1 & b_2 & \cdots & b_n \end{array} \right] \left[\text{of which } |a_i|^2 + |b_i|^2 = 1; i = 1, 2, \cdots, n \right] \quad (5.2)$$

It is also expressed as $A: \left| \overbrace{00\cdots0}^{n} \right\rangle$.

Through the transformation of Hadamard gate, S and A can be initialized, and when they are made to stay in superposed state:

$$S: H^{\otimes m} \left| \overbrace{00\cdots0}^{n} \right\rangle = \frac{1}{\sqrt{2^m}} \sum_{i=0}^{2^m - 1} |i\rangle \quad (5.3)$$

$$A: H^{\otimes n} \left| \overbrace{00\cdots0}^{n} \right\rangle = \frac{1}{\sqrt{2^n}} \sum_{i=0}^{2^n - 1} |i\rangle \quad (5.4)$$

The mapping from state to action is marked as:

$$f(s) = \pi : S \rightarrow A \tag{5.5}$$

Therefore, $f(s) = \left| a_s^n \right\rangle = \sum_{i=0}^{2^n-1} C_a \left| i \right\rangle$, of which $\left| C_a \right|^2$ stands for the occurrence probability of $\left| i \right\rangle$ when it is measured.

5.2.3 The Process of Quantum Reinforcement Learning

The process of quantum reinforcement learning is described as follows:

1. According to the expression (5.3) and (5.4), initialize the state set and action set of each agent, and assume the learning rate as γ.
2. Repeat the following procedures:
   a. Observe if there is a target for the next step of the state s and its surrounding world w, and if there is, a new round of learning is started, otherwise action a is chosen according to expression (5.5) and a certain exploration strategy is adopted. The Grover operator is used to update state s and world w, trying to keep the two consistent with each other.
   b. Executive action a. Observe the reward r and the next state s' and leave the corresponding footprints in the world at the same time. Update $V(s) = V(s) + \alpha \left[r + \gamma V(s') - V(s) \right]$ until it reaches $\left| \Delta V(s) \right| \leq \varepsilon$.

5.2.4 Sample Analysis

Design a 30 × 30 grid of the world for the agent that is a world of closed surface type, with no boundaries. So, when the agent moves to the far left, it doesn't reach the boundary, but appears in the far right, and the situation is the same in the far right, top, and bottom. Each agent can have eight actions in each position that are moving left, upper left, up, upper right, right, right down, down, left down, and its step length is 1. "Cover" the grid world by a "footprint world," and the so-called footprint world refers to that whenever the agent walks, it will leave a mark. If the next step is heading left, the corresponding "footprints" will deepen, fact that can provide references for another agent's decision-making.

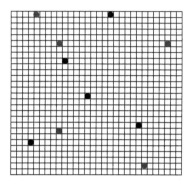

Figure 5.1 The grid world of the agent.

Each position of the grid world uses three qubits to express the state, whose amplitude probability of each superposed state stands for the depths of the footprints toward eight directions, respectively. The grid world of the agent is shown in Figure 5.1.

In Figure 5.1, a small solid black ball stands for the agent, and G for the goal. The behavior of each agent in any position is determined by strategy π, based on the current state. The multiagent system in this sample is of the cooperative type, so all the agents share the experience. By observing the footprints and their surrounding states, including whether there is an agent or a goal, they decide on the next action, according to their past experience.

Five agents and five goals have been put in the grid world. The learning rate $\gamma = 0.9$ is decreasing gradually, as the learning cycle increases. The task is accomplished when the agent finds any G, and the searching activities are underway at the same time. The results are shown in Figure 5.2.

5.3 THE DESIGN OF DISTRIBUTION NETWORK VULNERABILITY ANALYSIS SYSTEM BASED ON Q-MAS

The design purpose of the distribution network vulnerability analysis system based on Q-MAS is to make the system evaluate its vulnerability scientifically, and to take rapid and appropriate control measures to cope with system failures, and thus to ensure the reliability and security of the power system. Its main content includes: obtain real-time information widely from the changing environment, and

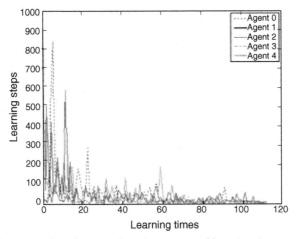

Figure 5.2 The comparison between learning steps and learning times.

explain them; be able to evaluate the vulnerability of the power system scientifically, and to limit the influence of faults in the smallest range. Based on the comprehensive evaluation of the whole system, establish an effective power protection mechanism to provide preventive and compensatory measures in order to improve the preparation of the power network.

5.3.1 The Overall Design

As the complexity of each electrical device in the distribution network and the composition of power plants increases, the requirements for control become increasingly stricter. If all the information collected from all the substations, users, and lines of the whole system are sent to the central dispatching center to be processed, the center then sending the processed information back to each substation, there will be a series of problems, such as a large amount of information, channel congestion, slow processing speed, complex operation, and so on; this is not feasible, both technically and economically. In view of this, we can monitor the large amount of dispersed information by adopting a hierarchical control structure, and setting up a central dispatching center, a substation control center, as well as a hierarchical control center of users/lines that will decompose the complex control of the whole distribution network into several interrelated control subsystems, decoupling the control of the multivariate process. Besides, we can set up an organizational layer agent in the central dispatching center, and build a coordinating-layer agent in the regional dispatching center and substation control center that is divided into line subsystems and user subsystems, according to the characteristics of the distribution network; meanwhile, the responsive-layer agent is set up in the subsystems and the logic hierarchical control structure is adopted in it, as shown in Figure 5.3. [6,7]

In physics, the system takes the agent coalition structure, of which the multiagent group is divided by geographical distance, having nothing to do with the agent's control layer. The bottom layer is the responsive layer (user/line), set up in each local subsystem to undertake predefined control behavior on the site. The middle layer is the coordinating layer (substation control center), and the coordinating-layer agent uses the knowledge base in order to diagnose the warnings or faults put forward by the responsive agent, and then send the control signal back to the responsive layer, coping with the faults in each subsystem in time, a fact that effectively limits the spread of faults. If the impact of the event exceeds some predefined limit, the

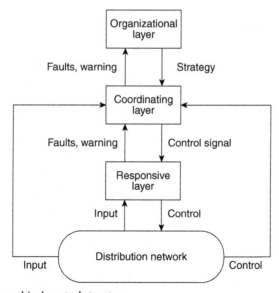

Figure 5.3 Hierarchical control structure.

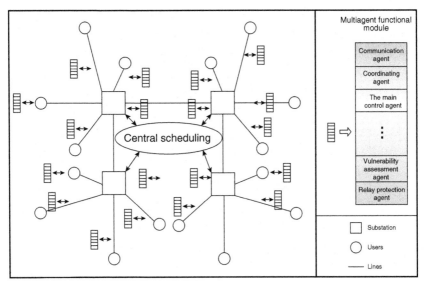

Figure 5.4 Schematic diagram of a vulnerability analysis system structure based on Q-MAS of the distribution network.

coordinating-layer agent will submit the event to the top layer, also known as the organizational layer (dispatching center).

Choose some important loads and lines from the distribution network that will affect the overall stability, combined with each substation and dispatching center, in order to get the main structure forming the analysis system. Set up a multiagent function module in each node, which has some similar control functions, and thus connect the system from the top to the bottom through a communication agent, in order to form a whole system that is able to control in a coordinated manner. Figure 5.4 shows a schematic diagram of a vulnerability analysis system structure of the distribution network based on Q-MAS.

5.3.2 The Design of Each Agent's Function in Multiagent Function Modules

The MAS function is very complex. When designing its components, we must understand fully the system's structure, and we need to define accurately the system behavior, in order to make the agent meet various needs. There are two types of agent that can be applied to the MAS of the power system, that is, proactive agent, and responsive agent. The proactive agent is a kind of intelligent agent, having its own knowledge base, able to cope with the coordination with other agent, as well as being able to make corresponding adjustment decisions according to the changes of environment. Yet, the responsive agent works in a simple way of stimulation-response.

As shown in Figure 5.5, this agent structure is applicable to the system. Protecting the responsive agent stands for the protection of the control devices (such as the relay) that undertake real-time control on the scene. The faults-isolation agent analyzes faults and disturbances in the environment, and sends the analysis results to the information-processing agent. The frequency-control agent monitors the frequency stability based on the preset frequency parameters determined by the mathematical relationship of frequencies with control and response based on frequencies. The information-processing agent processes faults alarm information coming from the bottom of the system, and generates important and valuable information that is sent to the fault-recognition agent. The fault-recognition agent will confirm the fault and its detailed information, based on the analysis results submitted by the information-processing agent. The model-updating agent updates the parameters in the current state model based on the data of real-time system, and checks whether the directive from the organizational layer is suitable for the current state of the system. The instruction-decomposition agent decomposes the information from the organizational level, and forms a more detailed directive to be transmitted to each designated agent. The vulnerability-assessment agent analyzes the system vulnerability constantly, and calculates the corresponding vulnerability index, as well as focusing on the monitoring results achieved by the hidden fault-monitoring agent, in order to find potential points of faults. The vulnerability-assessment agent is used for real-time monitoring of network communication, and in order to ensure the reliability of the communication system through the collaboration with the communication-service agent. When the communication system becomes weak, the communication-service agent will control the routes in the network in order to ensure that important instructions can be transmitted as a priority.

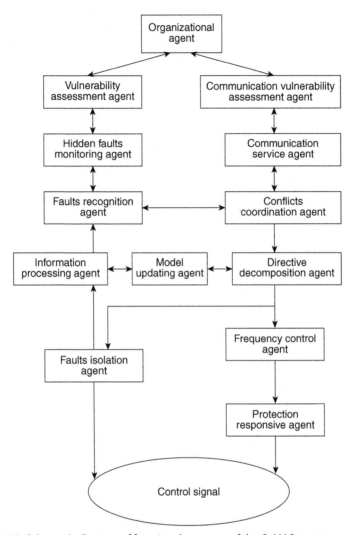

Figure 5.5 Schematic diagram of functional structure of the Q-MAS agent.

The hidden fault-monitoring agent monitors in real-time the common hidden faults in the power system (such as relay, generator set, etc.), and determines the scope of their location. The organizational agent belongs to the top agent, providing macro strategies for the whole system, including reference strategies and enforcement strategies. When conflicts occur among some agents during the operating process, the plan agent is responsible of coordination, in order to determine the best executive instructions.

Such a structure of Q-MAS is able to exert its organization and coordination ability so that it may evaluate system vulnerability, and develop a strategy of adaptive environment. The bottom layer responds quickly to system interference and faults and, meanwhile, the top layer analyzes the system from a global angle. If needed, the response of the bottom layer agent can be rejected by the top layer agent. For example, a protective agent from the responsive layer decides to trigger a relay from the perspective of its subsystem, but the organizational-layer agent determines from the global angle to stop triggering the relay, so the response from that certain protective agent will be rejected. Each agent in the system finishes its own task independently and, at the same time, the whole system can run steadily in coordinated interactions once again.

5.3.3 Communication Between Agents

As mentioned earlier, the system realizes the unified operation through collaboration between each agent, and communication is the foundation of interactive cooperation. The main communication ways of Q-MAS include request, inform, and promise, and so on; which one is chosen is determined by the task to be completed by the agents. Some communication examples of the system are used as an illustration.

1. Inform (protection-responsive agent → fault-isolation agent, "fault record"), the protection-responsive agent detects the system failure and determines the corresponding fault isolation control instruction, submitting the fault record to the faults-isolation agent.

2. Inform (information-processing agent → model-updating agent, "the current state of the system"), the model-updating agent uses the data submitted by the information-conversion agent to update in real-time the parameters of the system's current state.

3. Request (directive-decomposition agent → model-updating agent, "verify directive's validity") after the directive-decomposition agent receives a directive from the organizational layer, the

model-updating agent needs to confirm its validity. If the directive conflicts with the current state model, then the model-updating agent will inform the agent of the organizational layer to modify their decisions.

5.4 THE REALIZATION OF PRIMARY VULNERABILITY ASSESSMENT FUNCTION

The traditional security analysis method of the power system is deterministic security analysis [8–13]; this has been widely used in the power industry, providing a relatively high degree of reliability, but the results are too conservative because it only pays attention to the worst and most credible accidents. As such, the existing equipment has not been utilized fully from the perspective of operation, and some unnecessary redundant constructions are thus caused in the planning. In this case, the proposal of a safety assessment method based on risks has important practical significance that can get, quantitatively speaking, two factors that determine the level of security: the possibility and the severity of the accident, based on which, risk indicators are introduced in order to make the security assessment of the power system more scientific and detailed.

5.4.1 The Basic Principle of Risk Assessment

Risk theory is a theory combining the possibility that uncertainty factors may lead to disasters in the system, and their seriousness. Risk assessment refers to the fact that a series of logical steps are adopted to make the designers and safety engineers able to check, in a systematic way, the disasters caused by using devices, and to select the appropriate safety measures. The risk indicators of the power system can quantitatively grasp two factors that determine the system reliability: the possibility and the severity of the accident that can reflect comprehensively the influence of accidents on the whole power system, and the vulnerability degree of the corresponding system.

To realize the vulnerability assessment of the system, a risk assessment indicator is established to reflect the vulnerability degree of the distribution network that can get the factors determining the degree of safety quantitatively: the possibility and the seriousness of the accident. So, in the security assessment of the power system, risk is defined as: the probability of the accident and the result caused by it. Usually, we adopt expression (3.1) to calculate the risk value, which is:

$$R(C/X_t) = \sum_i P(E/X_t) \times S(C/E)$$

In this expression, the risk indicator $R(C/X)$ is an index of steady state vulnerability, reflecting the vulnerability degree of the power system; this is a comprehensive performance of the probability and seriousness of the accidents suffered by the power system. So, by calculating the risk indicator value of the system, the vulnerability state of the distribution network can be determined.

5.4.2 The Accident's Probability Model of Risk Assessment

In order to calculate the risk indicator of the power system, two parameters shall be obtained: the probability and the seriousness of the accident. The accident's probability refers to the occurrence possibility of the accident in the power system, including single-phase short circuit, three-phase short circuit, and open circuit, and so on. From the statistic data of historic accidents in the power system, as well as other references, it can be seen that the occurrence probability of an accident is in line with the Poisson distribution, that is:

$$P(E/X_t) = \frac{\lambda^k e^{-\lambda}}{k!} (k = 0,1,2\cdots)$$

In this expression, $P(E/X)$ is the probability that accident E has happened for k times, and λ is the mathematical expectation of occurring times of the accident in some period of time.

According to statistic materials, the scope of fault elements lies in the short-circuit faults of running lines in the power grid. The time unit of the fault rate is one year – and it only considers the faults of a single line, instead of the cases of parallel elements, or that two independent elements fail at the same time. The physical meaning of the accident rate achieved is the probability that each line fails one time within one year.

5.4.3 The Risk Indicator in Risk Assessment

As risk calculation has a strong decoupling feature, risks can be calculated according to each kind of security problems, each accident, or each element. Therefore, the overall risk assessment of the whole system can be divided into the evaluation of each kind of safety problems, and calculating the risk indicator value of different types, that can reflect different perspectives of the system security problems, as well as being able to achieve the target of evaluating the overall risk of the system affected by certain accidents. The risk assessment defines four types of risk-evaluation indicators. They are: line overload risk, transformer overload risk, low voltage risk, and loss of load risk.

1. Line overload risk indicator

    The line overload risk indicator reflects the comprehensive performance of the possibility and severity of transmission power overload of the normally running lines in the system, caused by some elements' faults. Define the severity of line overload risk as $S_L(C/E)$, and the power flow through the line determines the severity value of line overload risk. As shown in Figure 5.6, the line with real current that is lower or equal to half of the rated current belongs to the light load, whose severity value of overload risk is 0, as it satisfies $N - 1$ check of double circuit line; when the real current of the line is above or equal to 100% of its rated value, the line cannot operate normally, and its severity value of overload risk is 1; in the middle section, the severity function value of line overload risk is linear with the scale of the line load rate.

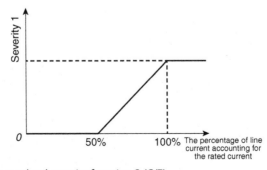

Figure 5.6 Line overload severity function $S_L(C/E)$.

Consequently, the calculation formula of the line overload risk indicator is:

$$R_L\left(C/X_t\right)=\sum_i P\left(E/X_t\right)\times S_L\left(C/E\right) \tag{5.6}$$

In this expression, $S_L(C/E)$ is the severity of overload risk in a certain running line; $R_L(C/X_t)$ is the total line overload risk of the whole system caused by some element fault.

2. Transformer overload risk indicator

The transformer overload risk indicator reflects the comprehensive performance of the possibility and the severity of large load rate or overload occurring in a normally operating transformer, caused by some system failures. The severity function of transformer overload risk is defined as $S_T(C/E)$, whose value is determined by the load rate of each main transformer. In Figure 5.7, when the transformer load rate is lower or equal to half of the rated capacity, it belongs to the light load, and if one of the two main transformers stops running, the other one can be fully loaded, so its severity value is 0; when the transformer load rate is higher or equal to 100%, its risk severity function value is 1; for the middle section, the severity value of transformer overload risk is linear with the size of its load rate.

Consequently, the calculation formula of the transformer overload risk indicator is:

$$R_T\left(C/X_t\right)=\sum_i P\left(E/X_t\right)\times S_T\left(C/E\right) \tag{5.7}$$

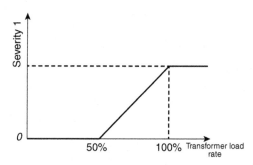

Figure 5.7 Transformer overload severity function $S_T(C/E)$.

In this expression, $S_T(C/E)$ is the overload severity of one transformer; $R_T(C/X_t)$ is the total transformer overload risk of the whole system caused by some element faults.

3. Low voltage risk indicator

Low voltage risk reflects the comprehensive performance of the possibility and severity of instances when the running bus voltage of the system drops to a very low level, instance caused by some accidents of the distribution network. In low voltage risk assessment, the severity function of low voltage risk of bus is defined as $S_V(C/E)$ and the voltage amplitude of each bus determines the value of its low voltage risk severity function. As shown in Figure 5.8, when the bus voltage is 1.0 (per-unit value), the voltage is normal, and its severity value is 0; when the bus voltage decreases to about 0.7 (per-unit value) or below, the load protection of the common motor will take effect and make it stop running, according to the load characteristics, so its value is 1; for the middle section, the value of low voltage risk severity function is linear with its bus voltage amplitude. In practice, over voltage problems rarely occur in the system, so when the bus voltage is higher than 1.0 (per-unit value), the values of low voltage risk severity functions are all 0.

Consequently, the calculation formula of the low voltage risk indicator is modified as:

$$R_V\left(C/X_t\right) = \sum_i P\left(E/X_t\right) \times S_V\left(C/E\right) \qquad (5.8)$$

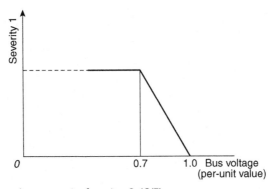

Figure 5.8 Low voltage severity function Sv(C/E).

In this expression, $S_V(C/E)$ is the low voltage severity of one bus that is caused by some accidents; $R_V(C/X_p)$ is the total low voltage risk of the whole system that is caused by some accidents.

4. Loss of load risk indicator

Loss of load risk reflects the comprehensive performance of the possibility and severity of loss of load that is caused by some accidents in the system. In loss of load risk assessment, the severity function of loss of load risk is $S_{load}(C/E)$. As shown in Figure 5.9, the severity of loss of load is set to be equal to the most serious cases of other safety problems, and it is constant with the value 1.0, no matter how great the loss of load is.

Consequently, the calculation formula of the low voltage and overload risk indicator is modified as:

$$R_{load}\left(C/X_t\right) = \sum_i P\left(E/X_t\right) \times S_{load}\left(C/E\right) \qquad (5.9)$$

In this expression, $S_{load}(C/E)$ is the loss of load severity caused by accidents; $R_{load}(C/X_p)$ is the loss of load risk of the whole system that is caused by some accidents.

5. Comprehensive risk indicator

The comprehensive risk indicator reflects the comprehensiveness of each kind of risks after the accidents, whose definition is the sum of four indicators, such as line overload risk indicator, transformer overload risk indicator, low voltage risk indicator, and loss of load risk indicator.

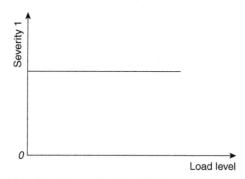

Figure 5.9 Severity function of loss of load $S_{load}(C/E)$.

Consequently, the calculation formula of the comprehensive risk indicator is:

$$R\left(C/X_t\right) = R_L\left(C/X_t\right) + R_T\left(C/X_t\right) + R_V\left(C/X_t\right) + R_{\text{load}}\left(C/X_t\right)$$

(5.10)

5.4.4 The Calculation Procedure of Risk Assessment

Four kinds of system security problems have been established in the assessment, describing the system vulnerability from different perspectives. Therefore, for each accident, the various types of risk indicators stated above should be calculated separately during a risk evaluation, as shown in Figure 5.10. In the calculation process, the target set of faults shall be determined first of all, including all the components with possible failures that need to be calculated. Then, known data are used to calculate the probability of each fault in the target set, as well as the power flow distribution of the system after each failure, based on which valid data from the power flow, results are applied into each risk model established above, in order to achieve various risk indicators under each fault. Finally, after all of the faults in the target set have been calculated, evaluate the system vulnerability through integrating these risk indicators.

5.5 THE REALIZATION OF COMPREHENSIVE VULNERABILITY ASSESSMENT FUNCTION OF THE DISTRIBUTION NETWORK

5.5.1 The Overall Design of Comprehensive Vulnerability Assessment of the Distribution Network

By calculating different indicator values, such as line overload risk, transformer overload risk, low voltage risk, and loss of load risk, and so on, the overall system risk assessment is decomposed into the single evaluation of various equipment security problems. But the distribution network is an organic whole, so the vulnerability of each level of equipment will inevitably affect the safety of other devices. For example, if line 1 is in a serious vulnerable working status, the working status of its adjacent line 2 is also of high vulnerability, even if line 2 has passed the single risk assessment. Therefore, during the assessment, devices of

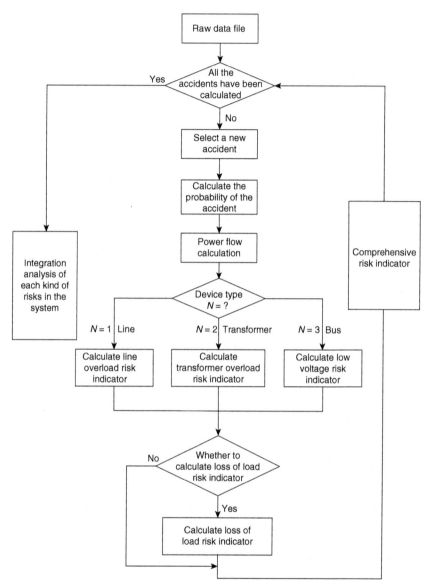

Figure 5.10 The calculation procedure of risk indicators.

different levels shall not be isolated, and should be seen as a whole; also, each level of lines and dispatching centers should take the influence of other related lines' or substations' vulnerability into account.

Based on this, a kind of assessment mechanism is designed with the evaluating level from low to high, from simple to complex, in order to analyze the comprehensive vulnerability of the overall distribution network, whose detailed assessment procedure is as follows:

1. First of all, according to different risk indicators, analyze the vulnerability of each level and each device separately, such as each line, or transformers of each substation.
2. Then, submit the vulnerability of low-level devices to the higher level that will calculate each device's vulnerability in the whole distribution network, through a comprehensive algorithm, considering the correlation of devices.
3. Finally, sort the devices of different levels and types in the distribution network, based on their comprehensive vulnerabilities, and give a final vulnerability analysis result.

Through the above design, it can be seen that, in order to get the accurate comprehensive vulnerability index, the key is to find a reasonable algorithm that is able to quantify accurately the influence coefficient affecting vulnerability among various devices. According to the long-term running experience of a power grid, vulnerability influential factors among different lines under various accident situations, or in various abnormal working states, are also different, and vulnerability correlation among this kind of lines or substations is hard to calculate by quantitative indicators; in practical terms, it depends on the field operators' experience to a large extent. Therefore, the experience of the operators is introduced to the comprehensive vulnerability assessment of the power grid by an analytical hierarchy process that turns qualitative analysis into a quantitative calculation to evaluate accurately the vulnerability of all the equipment in the distribution network.

5.5.2 Fuzzy Comprehensive Evaluation Decision Model Based on Analytical Hierarchy Process [14,15]

The analytical hierarchy process (AHP) is a multiobjective decision analysis method that combines qualitative and quantitative analysis, and is especially able to quantify the experience of policymakers, and is more practical in the case when the target (that is, factors) has a complex structure, yet lacks the necessary data.

1. The mathematical model of the analytical hierarchy process

   The analytic hierarchy process regards a complicated problem of multiobjective decision as an orderly hierarchical structure, and determines the relative weights of various factors in each layer based on the hierarchical structure diagram, using accumulated experience and considering the reality of problems to be solved until the relative weights of each plan at the bottom is worked out; this gives the pros and cons sequences of each plan. The detailed mathematical model is as follows:

a. Establish the problems' hierarchical structure models. When analyzing a problem, we should find out its range and clear all the factors contained in the problem, as well as their correlations, based on which we group these factors, according to whether they have something in common. Besides, their common elements are seen as a factor of the new level in the system that can, in itself, be integrated according to another set of characteristics in order to form a factor of higher layer until it becomes, in the end, a single factor of the highest level. Thus, a hierarchical structure model composed by the top level, by several middle layers, and the bottom layer is built up.

b. Construct a pairwise comparison judgment matrix. In the established hierarchical structure model, each layer consists of multiple elements, except the general objective layer, and the influence degree of various elements in the same layer on some certain element of the upper layer is different from each other; this compels us to judge the influence degrees that elements in the same layer have on some certain element of the upper layer, and quantify them. Constructing a pairwise comparison judgment matrix is a method to judge and quantify the influence degrees in the elements above.

c. Single-level sorting. A judgment matrix is the evaluation data of pairwise comparison that considers the upper layer, and single-level sorting is used to sort the priorities of all the elements in a level, compared to a certain element in the adjacent upper layer, also known as the eigenvector of the judgment matrix.

d. The total level of sorting. The total level of sorting is used to further calculate comprehensively the optimized order for the upper most layer (or general target layer) by using the calculation results of single-level sorting that is conducted layer by layer, from the top to the bottom.

2. The establishment of the decision model

In complex objective systems, because there are many factors to be considered, and each factor is placed on a different level – with most of them having strong fuzziness – it is difficult to sort or get meaningful decisions if only the fuzzy comprehensive evaluation model above is used. At this point, we need a multilevel fuzzy comprehensive evaluation model, whose basic idea is: first, evaluate each factor of the bottom layer comprehensively, and then move upwards and evaluate layer by layer, until the top layer is evaluated, in order to get the general assessment results. In this instance, the two–layer fuzzy comprehensive evaluation model is taken as an example to illustrate the decision-making process of the fuzzy comprehensive decision model that is based on an analytical hierarchical process.

U is the factor set, $V = \{v_1, v_2, \cdots, v_m\}$ is the remark set. Classify the factors in U into s groups, that is, $U = \{U_1, U_2, \cdots, U_s\}$, and it satisfies $U = \bigcup_{i=1}^{s} U_i$. Besides, when $i \neq j$, $U_i \cap U_j = \varnothing$. For each U_i, there is $U_i = \{U_{i1}, U_{i2}, \cdots, U_{in}\}$, of which n stands for the number of factors contained in the group i of the factor set. So, U is the top-level factor set, while U_i is the bottom level factor set.

$A_1, A_2, \cdots, A_s$ are the weight sets of factor sets $U_1, U_2, \cdots, U_s$ belonging to the second layer, and A is the weight set of the first layer factor set U, containing s factors; R_i represents the membership of the second layer factor set U_i with the remark set V, whose decision making process is as follows:

a. Make comprehensive decisions in the bottom layer, that is, $\forall i \in \{1, 2, \cdots, s\}$, and make comprehensive decisions on the decision space (U_i, V, R_i) in order to get the decision result $B_i = A_i \otimes R_i$.

b. Calculate the fuzzy decision matrix R of the high level composed by the comprehensive decision output B_i of the low level, that is:

$$R = \begin{bmatrix} B_1 \\ B_2 \\ \vdots \\ B_s \end{bmatrix} = \begin{bmatrix} A_1 \otimes R_1 \\ A_2 \otimes R_2 \\ \vdots \\ A_s \otimes R_s \end{bmatrix}_{s \times n} \qquad (5.11)$$

c. Make comprehensive decisions in the high level, which means to make integrated decisions on the decision space (U, V, R), of which $U = \{U_1, U_2, \cdots, U_s\}$, in order to get the comprehensive decision result B through the compositional operation $B = A \otimes R$.

This is the decision making process of the two-layer fuzzy comprehensive decision model. For the multilayer fuzzy comprehensive decision making process, the process is similar with the above. Only by executing repeatedly step b. and c. on the middle layer, until it reaches the top level, the final decision result can be gained. Classify each device in this project and judge the influence degree that can be divided into severe impact, slight impact, and no impact, according to factors such as reliability, economy, and maintainability of the devices. Because the indicators measuring the importance of the equipment are the ones with fuzziness from many kinds of perspectives, these factors have priority problems, namely the weighting problems among factors. Therefore, adopting a fuzzy mathematical method to make decisions on the maintenance mode of devices is reasonable and feasible.

3. Set up the index system

In line with the principle of refining, a vulnerability influence index system is established according to a fuzzy comprehensive evaluation method that shows scientificity and pertinence, featuring the characteristics of each perspective of the devices, so that this index system is also very concise and practical. The project described in this chapter has designed a set of two-level evaluation index system that calculates the vulnerability of a device and the vulnerability influence degree of other related devices, as shown in Figure 5.11.

The first class is about the evaluation aspect, including nonfault indicators and fault indicators, of which the nonfault indicator refers to the vulnerability of related devices that is caused by the device's abnormal operation, and the fault indicator refers to the vulnerability of related devices that is caused by device's fault state. Seven indicators are derived from the

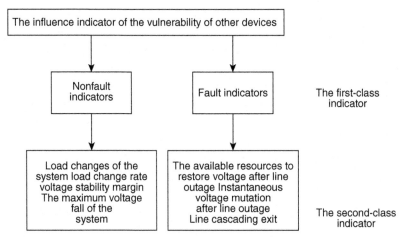

Figure 5.11 The evaluation index system of device maintenance mode.

two indexes that elaborate various influences. The evaluation index system $U = \{U_1 (\text{nonfault indicators}), U_2 (\text{fault indicators})\}$ has $U_1 = \{u_{11}, u_{12}, u_{13}, u_{14}\}$ (four factors) and $U_2 = \{u_{21}, u_{22}, u_{23}\}$ (three factors).

4. Set up the remark set

The remark set is a set composed by various decision results probably made by policy makers on the evaluation objective. The objective of fuzzy comprehensive decision-making is to achieve the best decision result from the remark set, on the basis of considering all the factors comprehensively. The remark set can be determined by the qualitative evaluation method, and by the quantitative rating method that is achieved through the experts' experience and reasoning, combined with the instance.

Divide the remark set into server impact, light impact, and no impact, that is, $V = \{v_1, v_2, \cdots, v_m\}$, of which v_1 is server impact, v_2 light impact, v_3 no impact.

5. Determine the fuzzy decision matrix

Determine the relationship matrix $R_i = \left(r_{ij} \right)_{mn}$ of each evaluation factor set u_i with remark set V, and each decision factor used in this project is able to compose the fuzzy decision matrix below:

$$R_1 = \begin{bmatrix} r_{11} & r_{12} & r_{13} \\ r_{21} & r_{22} & r_{23} \\ r_{31} & r_{32} & r_{33} \\ r_{41} & r_{42} & r_{43} \\ r_{51} & r_{52} & r_{53} \\ r_{61} & r_{62} & r_{63} \\ r_{71} & r_{72} & r_{73} \end{bmatrix} \tag{5.12}$$

In this expression, r_{ij} is the membership of evaluation factor u_i to evaluation grade v_j.

6. Determine the result based on fuzzy comprehensive decision-making

In 1965, the American professor L. A. Zadeh created the fuzzy set theory, describing the middle transition of differences by the degree of membership, that is, a description of fuzziness using accurate mathematical language.

We have got the index set U to evaluate device's importance, recorded as:

$$U = \left(U_1, U_2, \cdots, U_n \right) \tag{5.13}$$

Different weights are set to each indicator U_i in the set U, according to its impact degree on the device's importance, forming a weight set A of evaluation indicators, recorded as:

$$A = \left(A_1, A_2, \cdots, A_i, \cdots, A_n \right) \left(A_i \geq 0, \sum_{i=1}^{n} A_i = 1 \right) \tag{5.14}$$

A is the fuzzy subset of U.

The remark set is marked as:

$$V = \left(V_1, V_2, \cdots, V_m \right) \tag{5.15}$$

The operating staff should determine the membership grade R of the influence degree of device vulnerability on related device vulnerability to each remark, on each evaluation indicator. The membership grade of indicator i to each remark level is the fuzzy subset of remark set B.

In single index evaluation, it is recorded as:

$$R_i = \left(r_{i1}, r_{i2}, \cdots, r_{im} \right) \tag{5.16}$$

In multi–index evaluation, the comprehensive evaluation matrix R is:

$$R = \begin{pmatrix} r_{11} & r_{12} & \cdots & r_{1m} \\ r_{21} & r_{22} & \cdots & r_{2m} \\ \vdots & \vdots & \vdots & \vdots \\ r_{n1} & r_{n2} & \cdots & r_{nm} \end{pmatrix} \tag{5.17}$$

After taking evaluation indicators weights into account, the evaluation matrix B of each device given by the operating staff can be gained:

$$B = A \otimes R = \left(A_1, A_2, \cdots, A_i, \cdots, A_n \right) \otimes \begin{pmatrix} r_{11} & r_{12} & \cdots & r_{1m} \\ r_{21} & r_{22} & \cdots & r_{2m} \\ \vdots & \vdots & \vdots & \vdots \\ r_{n1} & r_{n2} & \cdots & r_{nm} \end{pmatrix} \tag{5.18}$$

$$= \left(B_1, B_2, \cdots, B_j, \cdots, B_m \right)$$

In the expression, i is the synthetic operation symbol; B_j is the membership grade of influence degree j of the device vulnerability to the other device vulnerability evaluated by an operator.

According to the maximum membership principle, if

$$B_k = \max \left(B_1, B_2, \cdots, B_m \right) \tag{5.19}$$

The influence degree of this device vulnerability on the other device vulnerability is modified as level k.

7. The decision–making examples of two related lines vulnerability influence degree

Take wheel link line A and B of a domestic oilfield distribution network, for example, and carry out a decision-making process about the impact degree of a device's vulnerability to another device's vulnerability. Its network structure is shown in Figure 5.12, in which two lines bear the power supply of six loads at the same time. Now, take wheel link line A's vulnerability influence on wheel link line B's vulnerability as an example to illustrate how to calculate the influence degree of any device's vulnerability on another device's vulnerability.

a. Set up the index system. According to the description mentioned above, the problems can be divided into two layers. The first layer includes nonfault indicators and fault indicators, expressed as $U = \{U_1, U_2\}$. The second layer contains seven factors, including the maximum voltage fall of the system, voltage stability margin, load changing rate, load changes of the system, line cascading exit, instantaneous voltage mutation after line outage, and feasible resources to restore voltage after the

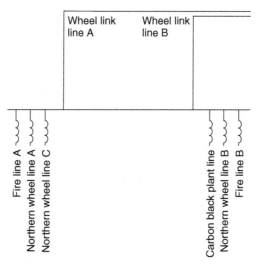

Figure 5.12 Network structure of round link line A and round link line B in distribution network of some oil field.

line outage, expressed as $U_1 = \{u_{11}, u_{12}, u_{13}, u_{14}\}$ (four factors) and $U_2 = \{u_{21}, u_{22}, u_{23}\}$ (three factors).

b. Build the remark set. Here, we sort the remark set into severe impact, light impact, and no impact, which is expressed as $V = \{v_1, v_2, \cdots, v_m\}$, v_1 being the severe impact, v_2 the light impact, v_3 having no impact.

c. Determine weighting factors. We need to put different coefficients to each evaluation indicator, since its effect on the evaluation result is different, in order to get the result that can best meet the practical situation.

- Determine the weights of two main indicators: nonfault indicators, and fault indicators. Get the weight vector matrix according to the pairwise comparison of each indicator's importance. The comparison between the maintenance difficulty degree and two indicators of spare part supply are in Table 5.1. Therefore, the following judgment matrix is achieved.

$$A = \begin{bmatrix} 1 & 3 \\ \dfrac{1}{3} & 1 \end{bmatrix}$$

The square root method is used to calculate the priority vector:

i. Calculate the geometric means of all the elements in each row of the judgment matrix.

Table 5.1 Comparison of maintenance difficulty degree and two indicators of spare part supply.

Scale a_{ij}	Definition
1	Each factor is of the same importance with itself
3	Compared to fault indicators, nonfault indicators are slightly more important
Reciprocal	If comparing factor i with factor j, we can get judgment value $a_{ij} = \dfrac{1}{a_{ji}}$, $a_{ii} = 1$

$$\tilde{\omega}_1 = \sqrt{\sum_{i=1}^{n} a_i} = \sqrt{1+3} = \sqrt{4} = 2$$

$$\tilde{\omega}_2 = \sqrt{\sum_{i=1}^{n} a_i} = \sqrt{\frac{1}{3}+1} = \sqrt{\frac{4}{3}} \approx 1.1547$$

So we get:

$$\tilde{\omega} = \begin{bmatrix} 2, & 1.1547 \end{bmatrix}^T$$

ii. $\tilde{\omega}$ is normalized to get:

$$\omega_1 = \frac{\tilde{\omega}_1}{\sum_{j=1}^{n} \omega_j} = \frac{2}{3.1547} \approx 0.6400$$

$$\omega_2 = \frac{\tilde{\omega}_2}{\sum_{j=1}^{n} \tilde{\omega}_j} = \frac{1.1547}{3.1547} \approx 0.3600$$

So, after normalization, there is:

$$\omega = \begin{bmatrix} 0.64, & 0.36 \end{bmatrix}^T$$

- The calculation of weight coefficients of four third-level indicators of nonfault indicators. According to the pairwise comparison between the importance of each index factor, the following judgment matrix can be achieved:

$$A = \begin{bmatrix} 1 & \dfrac{1}{2} & 3 & 7 \\ 2 & 1 & 6 & 8 \\ \dfrac{1}{3} & \dfrac{1}{6} & 1 & 3 \\ \dfrac{1}{7} & \dfrac{1}{8} & \dfrac{1}{3} & 1 \end{bmatrix}$$

The square root method is used to calculate the priority vector:

i. Calculate the geometric means of all the elements in each row of the judgment matrix.

$$\tilde{\omega}_{21} = \sqrt{\sum_{j=1}^{n} a_{1j}} = \sqrt{1 + \frac{1}{2} + 3 + 7} = \sqrt{11.5} \approx 3.4$$

$$\tilde{\omega}_{22} = \sqrt{\sum_{j=1}^{n} a_{2j}} = \sqrt{2 + 1 + 6 + 8} = \sqrt{17} \approx 4.1$$

$$\tilde{\omega}_{23} = \sqrt{\sum_{j=1}^{n} a_{3j}} = \sqrt{\frac{1}{3} + \frac{1}{6} + 1 + 3} = \sqrt{\frac{81}{18}} \approx 2.12$$

$$\tilde{\omega}_{24} = \sqrt{\sum_{j=1}^{n} a_{4j}} = \sqrt{\frac{1}{7} + \frac{1}{8} + \frac{1}{3} + 1} = \sqrt{\frac{186}{168}} \approx 1.265$$

So we get:

$$\tilde{\omega}_2 = \begin{bmatrix} 3.4, & 4.1, & 2.12, & 1.265 \end{bmatrix}^T$$

ii. Normalize $\tilde{\omega}_2$ to calculate:

$$\omega_{21} = \frac{\tilde{\omega}_{21}}{\sum_{j=1}^{n} \tilde{\omega}_{2j}} = \frac{3.4}{10.885} \approx 0.3$$

$$\omega_{22} = \frac{\tilde{\omega}_{22}}{\sum_{j=1}^{n} \tilde{\omega}_{2j}} = \frac{4.1}{10.885} \approx 0.4$$

$$\omega_{23} = \frac{\tilde{\omega}_{23}}{\sum_{j=1}^{n} \tilde{\omega}_{2j}} = \frac{2.12}{10.885} \approx 0.2$$

$$\omega_{24} = \frac{\tilde{\omega}_{24}}{\sum_{j=1}^{n} \tilde{\omega}_{2j}} = \frac{1.265}{10.885} \approx 0.1$$

Finally, we get:

$$\omega_2 = \begin{bmatrix} 0.3, & 0.4, & 0.2, & 0.1 \end{bmatrix}^T$$

Take 0.64, the weight coefficient of product quality itself into account, and we achieve:

$$\omega'_{21} = 0.3 \times 0.64 = 0.192$$
$$\omega'_{22} = 0.4 \times 0.64 = 0.256$$
$$\omega'_{23} = 0.2 \times 0.64 = 0.128$$
$$\omega'_{24} = 0.1 \times 0.64 = 0.064$$

So, the coefficients of these four indicators weights in the whole judgment process are:

$$\omega'_2 = \begin{bmatrix} 0.192, & 0.256, & 0.128, & 0.064 \end{bmatrix}^T$$

- Determine the weight coefficient of each indicator of fault index. According to the pairwise comparison of the importance of each indicator factor, the weight vector matrix is achieved.

$$A = \begin{bmatrix} 1 & 6 & 5 \\ \dfrac{1}{6} & 1 & \dfrac{1}{3} \\ \dfrac{1}{5} & 3 & 1 \end{bmatrix}$$

The square root method is used to calculate the priority vector:

i. Calculate the geometric means of all the elements in each row of the judgment matrix.

$$\tilde{\omega}_{31} = \sqrt{\sum_{i=1}^{n} a_{3j}} = \sqrt{1+6+5} = \sqrt{12} \approx 3.46$$

$$\tilde{\omega}_{32} = \sqrt{\sum_{i=1}^{n} a_{3j}} = \sqrt{\frac{1}{6}+1+\frac{1}{5}} = \sqrt{\frac{41}{30}} \approx 1.2247$$

$$\tilde{\omega}_{33} = \sqrt{\sum_{j=1}^{n} a_{3j}} = \sqrt{3+\frac{1}{5}+1} = \sqrt{4.2} \approx 2.049$$

Finally, we get:

$$\tilde{\omega}_3 = \begin{bmatrix} 3.46, & 1.2247, & 2.049 \end{bmatrix}^T$$

ii. Normalize $\tilde{\omega}_3$ to calculate:

$$\omega_{31} = \frac{\tilde{\omega}_{31}}{\sum\limits_{j=1}^{n} \tilde{\omega}_{3j}} = \frac{3.46}{6.7337} \approx 0.52$$

$$\omega_{32} = \frac{\tilde{\omega}_{32}}{\sum\limits_{j=1}^{n} \tilde{\omega}_{3j}} = \frac{1.2247}{6.7337} \approx 0.18$$

$$\omega_{33} = \frac{\tilde{\omega}_{33}}{\sum\limits_{j=1}^{n} \tilde{\omega}_{3j}} = \frac{2.049}{6.7337} \approx 0.3$$

After normalization we get:

$$\omega_3 = \begin{bmatrix} 0.52, & 0.18, & 0.3 \end{bmatrix}^T$$

Take 0.36, the weight coefficient of produce quality itself into account, we get:

$$\omega'_{31} = 0.52 \times 0.36 = 0.1872$$
$$\omega'_{32} = 0.18 \times 0.36 = 0.0648$$
$$\omega'_{33} = 0.3 \times 0.36 = 0.108$$

So the weight coefficients of these three indicators among the whole judgment process are:

$$\omega'_3 = \begin{bmatrix} 0.1872, & 0.0648, & 0.108 \end{bmatrix}$$

From the above, the weight coefficients of each indicator among decision making process can be gained, and these are the influence on other devices in the assessment system, the impact on safety production, device qualities, special degree, the values of devices, maintenance cost, repair costs, the difficulty degree of maintenance and spare parts supply.

$$A = \begin{bmatrix} 0.192, & 0.256, & 0.128, & 0.064, & 0.1872, & 0.0648, & 0.108 \end{bmatrix}$$

d. Decision making results. Regard fifteen operational personnel as the evaluation group, evaluating each influential factor to get fuzzy evaluation matrix:

$$R = \begin{bmatrix} 0.334 & 0.333 & 0.333 \\ 0.133 & 0.534 & 0.333 \\ 0.289 & 0.489 & 0.222 \\ 0.489 & 0.289 & 0.222 \\ 0.334 & 0.533 & 0.133 \\ 0.511 & 0.117 & 0.312 \\ 0.267 & 0.401 & 0.332 \end{bmatrix}$$

Weight coefficients are:

$$A = \begin{bmatrix} 0.192, & 0.256, & 0.128, & 0.064, & 0.1872, & 0.0648, & 0.108 \end{bmatrix}$$

From the above expression, the comprehensive evaluating result of this device can be gained, as:

$$B = \left(b_1, \ b_2, \ b_3 \right) = A \otimes R \tag{5.20}$$

Of which b_1, b_2, b_3 are the proportions of severe impact, slight impact, and no impact after comprehensive evaluation, and common matrix operation is adopted in compositional operation $\otimes$ here to get $B = \begin{bmatrix} 0.29086, & 0.43627, & 0.2727 \end{bmatrix}$.

From the calculation of the expression above, it can be seen that the weight of slight impact is the greatest, so the influence of round link line A's vulnerability on round link line B's vulnerability is of slight degree.

From the network structure, we can also get that round line A and B supply loads for six lines together, so when crisis occurs in round link line A, it will surely affect the working status of round link line B, a fact that will further increase its vulnerability.

8. Assessment on comprehensive vulnerability of each device in the distribution network

Through the mathematical operation above, the influence degree of other related devices on one device can be achieved, and a result of comprehensive vulnerability assessment considering other related devices influence, and its own operational status, will also be received, if combined with the vulnerability indicators achieved through single device's vulnerability assessment, so any device's comprehensive vulnerability result can be calculated through the following expression:

$$R_i = R_i\left(C/X_t\right) + \sum_{n=1,n \neq i}^{k} w_{ni} R_n\left(C/X_t\right) \qquad (5.21)$$

In this expression, $R_i(C/X_t)$ is the vulnerability value of device i achieved from risk assessment; $R_n(C/X_t)$ is the vulnerability value of device n achieved from risk assessment; w_{ni} is the influence degree of device n's vulnerability on device i's vulnerability that is $w_{ni} = [\text{severe impact}, \quad \text{slight impact}, \quad \text{no impact}]$, and in this project it is quantified as $w_{ni} = (0.15, \quad 0.08, \quad 0)$; R_i is the comprehensive vulnerability value of device i. Sort R_i values of all the devices and the final vulnerability sequence will be achieved.

BIBLIOGRAPHY

[1] W. Chen, Q. Jiang, Y. Cao, et al. Risk assessment of voltage collapse in power system, Power Syst. Technol. 29 (19) (2005) 6–11.

[2] X. Wang, T. Zhu, P. Xiong, Vulnerability assessment and control of power system based on MAS, J. Electr. Power Syst. Autom. 15 (3) (2003) 20–22.

[3] C.J.C.H. Watkins, P. Dayan, Q-learning, Mach Learn 8 (3/4) (1992) 279–292.

[4] J. Hu, M.P. Wellman, Multi-agent reinforcement learning: Theoretical framework and an algorithm, in: Proceedings of the Fifteenth International Conference on Machine Learning, Morgan Kaufmann Publishers Inc., San Francisco, CA, 1998, pp. 242–250.

[5] Grover L.K., A fast quantum mechanical algorithm for database search, in Proceedings of the twenty-eighth annual ACM symposium on the theory of computing, Philadelphia, PA, May 22–24, 1996 (New York: ACM 1996) 212–219; Grover LK, Quantum mechanics helps in searching for a needle in a haystack, Phys Rev Lett 1997, 79: 325–328.

[6] W. Hua, J.D. McCalley, V. Vittal, et al. Risk based voltage security assessment, IEEE Trans. Power Syst. 15 (4) (2000) 1247–1254.

[7] C. Wang, Y. Xuyang, Distributed coordination emergency control of multiagent system, Power Syst. Technol. 28 (3) (2004) 1–5.

[8] W. Chen, Q. Jiang, Y. Cao, Voltage vulnerability assessment based on risk theory and fuzzy inference, Proc. Chin. Soc. Electr. Eng. 25 (24) (2005) 20–25.

[9] E. Fu, Economic management of devices, China Railway Publishing House, Beijing, (1994).

[10] A. Sarwat, Optimierung der anlageninstandhaltung, Erich Schmidt Verlag (1989) 116–119.

[11] M. Wolfgang, Integrierte anlagenwirtschaft, Verlag TUV Rheinland (1988) 72–73.

[12] G. Levitin, A. Lisnianski, Optimization of imperfect preventive maintenance for multistate systems, Reliab. Eng. Syst. Safe. (2) (2000) 193–203.

[13] Y.X. Zhao, On preventive maintenance policy of a critical reliability level for system subject to degradation, Reliab. Eng. Syst. Safe. (3) (2003) 301–308.

[14] J. Zhang, Control theory and engineering applications of fuzzy neural network, Harbin Institute of Technology Press, Harbin, (2004).

[15] Z. Liu, Y. Liu, Theory research and exploration of fuzzy logic and neural network, Beijing University of Aeronautics and Astronautics Press, Beijing, (1996).

CHAPTER 6

Low-Voltage Risk Assessment of the Distribution Network Based on Vulnerability Information Coordination Among Buses

Contents

6.1 INTRODUCTION

In recent years, the increasing complexity of power grid structures has gradually extended the operating vulnerability of the system. Moreover, as an important part of power supply and distribution, the distribution network is located at the end of the power system, connected directly to the user, whose stability reflects directly the quality of the power supply [1–3]. In addition, the low–voltage operation of the power grid will lead to problems such as dim lighting, increasing line

X. Meng and Z. Pian: Intelligent Coordinated Control of Complex Uncertain Systems for Power Distribution Network Reliability. http://dx.doi.org/10.1016/B978-0-12-849896-5.00006-4

power losses and decreasing the communication quality of motors and other equipment that may even destroy electrical equipment, in some serious cases, resulting in system collapse. Substantial growth in the load of the distribution network, as well as its own power supply radius, have also made, in recent years, the bus voltage of the distribution network goes very often below the limit value [4,5], but the issue only appears when a failure occurs; this has brought a significant security risk to the system operation. As such, the study of the low-voltage risk problem of the distribution network is of great significance.

Traditional risk theory overcomes the shortcomings of deterministic analysis and probabilistic analysis, taking into account the uncertainty and seriousness of power grid accidents. The traditional system vulnerability assessment algorithm often isolates each node in the evaluation system, solely neglecting the interaction among adjacent nodes' operations. Because of this problem, this chapter introduces the analytical hierarchy process (AHP) and fuzzy evaluation method in order to quantify the evaluation of field staff, and to calculate the interrelationship of vulnerabilities among nodes, based on which a comprehensive risk evaluation indicator that is in line with engineering practice is also proposed, combined with the risk assessment theory needed to calculate the overall risk of nodes. Finally, the simulation result of the proposed method is compared to that of the traditional method to demonstrate the effectiveness and practicality of the method proposed.

6.2 LOW-VOLTAGE RISK ANALYSIS OF ISOLATED BUSES

6.2.1 Low-Voltage Vulnerability Assessment of Isolated Buses Based on Risk Theory

Risk theory considers the randomness factor of the system that will lead to the possibility of disasters, combined with the severity of this kind of disaster. Risk is usually defined as "reflecting the possibility of disaster and its severity." This demonstrates that the possibility of accidents and the seriousness of the consequences are two important factors of risk. Accordingly, the calculation formula of the risk indicator is achieved as shown in equation (6.1).

When quantitative risk indicators are used to measure the vulnerability of a bus node, the greater risk value is, the more vulnerable the corresponding bus node is; otherwise, it is relatively strong. Risk is a bridge connecting safety and economy, local and global, as well as time point and the period of time.

1. The possibility of fault in the distribution network
   Calculating risk indicators need to calculate two parameters of an accident: possibility and severity. Power grid accidents mainly include short circuit and line disconnected. Drawn from statistical data and references, probabilities of power grid accidents are basically in line with the Poisson distribution [6], that is:

   $$P(E/X_t) = \frac{\lambda^k e^{-\lambda}}{k!}(k = 0,1,2\cdots) \tag{6.1}$$

   In this expression, $P(E/X_t)$ is the probability of accident E happening for k times; λ is the mathematical expectation of occurrence times of an accident in a certain period of time.

2. Severity of the distribution network faults
   Assume the low voltage severity as 0 when bus voltage is above its rated value, excluding the voltage influence. According to the voltage distortion rate formulated by the industry, it is generally believed that when the system voltage is equal to 0.95 (per unit), the severity of low voltage is defined as 1, and when it is less than 0.95, the severity is greater than 1, as shown in shaded portion of Figure 6.1. Besides, the severity of a power grid accident is also connected with the initial state, operating condition, and status of the power grid, and so on. Based on the experience and information available, the severity function of bus low voltage is defined as shown in Figure 6.1.

3. Low-voltage risk indicators of isolated buses
   Risk assessment can quantify the possibility and severity of accidents, connecting economy with security. Based on the earlier description, the quantitative expression of the low-voltage risk indicator of an isolated bus is:

   $$R_V(C/X_t) = P(E/X_t) \times S_V(C/E) \tag{6.2}$$

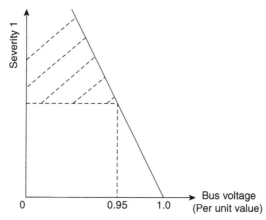

Figure 6.1 Severity function of bus low voltage in the distribution network.

In the expression, $S_V(C/E)$ is the low-voltage severity of one bus caused by accidents; $R_V(C/X_l)$ is the low-voltage risk value of one bus caused by some accidents.

6.2.2 Calculation Procedure of Low-Voltage Risk Indicator Value

The calculation steps of risk indicator value are: set up the expected fault set; calculate the possibility of each accident in the expected fault set; calculate the power flow distribution situation of the power grid after each failure, and if the node voltage in power flow result is less than 0.95, there is a low voltage risk, that is, calculate the indicator's value of bus low-voltage when the node voltage is lower than 0.95. The process for the calculation of risk indicator value is shown in Figure 6.2.

6.3 QUANTITATIVE CALCULATION OF INTERRELATIONSHIP OF VULNERABILITIES BETWEEN BUSES

The first section has achieved the single risk indicator value of each node and, subsequently, the AHP and fuzzy method will be integrated in order to calculate the mutual influence coefficient ω among bus nodes.

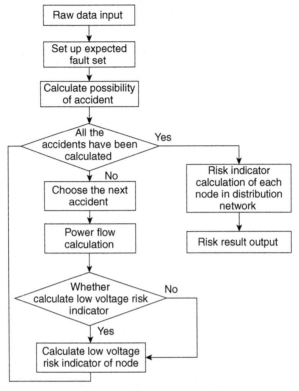

Figure 6.2 The calculation procedure of risk indicator value.

6.3.1 Fuzzy Analytical Hierarchy Process

AHP handles the multi-indicator issue by establishing a hierarchical structure, quantifying the hierarchical decision-making process in accordance with the laws of thinking and psychology, by reasonably combining the qualitative and quantitative decisions [7,8]. Fuzzy comprehensive evaluation is a very effective multifactor decision-making method that works by overall evaluation of things affected by a variety of factors, and whose feature is that its evaluation results are not absolutely positive or negative, but are expressed in a fuzzy set [9,10]. Hereby, AHP and fuzzy comprehensive evaluation is integrated in order to judge the influence of vulnerabilities among buses.

6.3.2 Indicator Selection Based on the Mutual Influence of Low-Voltage Vulnerabilities between Buses

In order to determine quantitatively the size of mutual influence of low-voltage vulnerabilities between buses, a series of indicators reflecting accurately the influence of low-voltage vulnerability among buses should be found, and it should follow the principles of scientificity, practicability, wide application and predictability, and so on. Guided by these principles, and based on the experience of onsite operating personnel, normal operation and abnormal operation indicators are chosen as the main factors affecting voltage stability [11–14].

1. Normal operating index set U_1

   $U_1 = \{u_{11}, u_{12}, u_{13}, u_{14}\}$ = {load level indicator, load rising rate indicator, maximum voltage exceeding lower limit indicator, maximum voltage fall indicator of the system}.

   u_{11} and u_{12}: During normal operation, if the load increases very fast, or by a great amount, it will make the voltage drop dramatically.

   u_{13} and u_{14}: If the voltage exceeds the maximum lower limit with no measures adopted, it will lead to voltage collapse. Reactive power plays a leading role in voltage stability, and if reactive power is not planned reasonably, it will lead to large voltage drop that will cause voltage collapse in this situation, if disturbance occurs.

2. Abnormal operation index set U_2

   $U_2 = \{u_{21}, u_{22}, u_{23}\}$ = {instantaneous mutation voltage indicator upon line-outage, indicator of overall reactive unbalance after line-outage, indicator of reactive unbalance covering the largest area after line-outage}.

   u_{21}: The lowest voltage value in the fault area can be calculated when a fault occurs in the distribution network; it is of great importance to low-voltage vulnerability assessment.

   u_{22}: If system reactive unbalance is in serious shortage after a fault occurs, it will affect the voltage stability seriously.

   u_{23}: Regional reactive unbalance may sensitively reflect the weakness of grid voltage, so this indicator is of great significance to mutual influence of bus vulnerability.

The indicators mentioned can be achieved by calculation, measurement, or empirical data.

6.3.3 Calculate the Mutual Influence Coefficient of Bus Vulnerability by Integrated Fuzzy AHP

1. Index system

   The low–voltage index system of mutual influence between buses in a distribution network is shown in Figure 6.3.

2. Set up the reviews set

   Here, the reviews set will be divided into server impact, light impact, and no impact, that is, $V = \{v_1, v_2, v_3\}$, v_1 is severe impact, v_2 is slight impact, and v_3 is no impact.

3. Determine the fuzzy decision matrix

   Determine the relation matrix $R_i = \left(r_{ij} \right)_{mn}$ of each evaluation factor set u_i to reviews set V, and each decision factor can make up the following fuzzy decision matrix:

$$
R_1 = \begin{bmatrix} r_{11} & r_{21} & r_{31} & r_{41} & r_{51} & r_{61} & r_{71} \\ r_{12} & r_{22} & r_{32} & r_{42} & r_{52} & r_{62} & r_{72} \\ r_{13} & r_{23} & r_{33} & r_{43} & r_{53} & r_{63} & r_{73} \end{bmatrix} \tag{6.3}
$$

   In the expression, r_{ij} is the membership of evaluation factor u_i to evaluation level v_j.

4. Determine conclusions based on fuzzy comprehensive decision

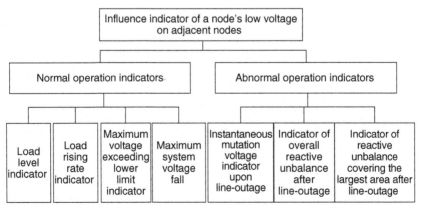

Figure 6.3 Low-voltage index system of mutual influence between buses of the distribution network.

The evaluation indicator set U of bus importance is known, recorded as:

$$U = \left(U_1, U_2, \cdots, U_n\right)$$

In the set U, each indicator U_i is weighted differently according to its different influence on device in order to form a weight set A of evaluation indicators, and A is a fuzzy subset of U, recorded as:

$$A = \left(A_1, A_2, \cdots, A_i, \cdots, A_n\right)\left(A_i \geq 0, \sum_{i=1}^{n} A_i = 1\right) \tag{6.4}$$

Field personnel determine the membership grade R of the influence degree of one bus vulnerability on other related bus vulnerability to each remark of each evaluation indicator. The membership grade of indicator i on each remark level is a fuzzy subset of reviews set B.

During multi-indicator assessment, the comprehensive evaluation matrix R is:

$$R = \begin{pmatrix} r_{11} & r_{12} & \cdots & r_{1m} \\ r_{21} & r_{22} & \cdots & r_{2m} \\ \vdots & \vdots & \vdots & \vdots \\ r_{n1} & r_{n2} & \cdots & r_{nm} \end{pmatrix} \tag{6.5}$$

Take into account the weights of evaluation indicator to obtain evaluation matrix B of operating personnel on each bus.

$$B = A \otimes R = \left(B_1, B_2, \cdots, B_j, \cdots, B_m\right) \tag{6.6}$$

In the expression, $\otimes$ is a compositional operation symbol; B_j is the membership grade when operating personnel rate the influence degree of vulnerability of this bus on that of the other one as level j.

According to the maximum membership principle, if

$$B_k = \max\left(B_1, B_2, \cdots, B_m\right) \tag{6.7}$$

The influence degree of vulnerability of this bus on that of the other one is rated as level k, so influence coefficient ω of vulnerability of one bus on that of the other one is the maximum value in matrix B.

6.4 COMPREHENSIVE INDICATOR OF LOW-VOLTAGE VULNERABILITY IN THE DISTRIBUTION NETWORK

The expression of bus comprehensive vulnerability in the distribution network is achieved through the following description:

$$R_I = R_i + \sum_{n=1,n\neq i}^{k} \omega_{ni} R_n \tag{6.8}$$

In the expression, R_i is the vulnerability value of bus i achieved through risk assessment; R_n is the vulnerability value of bus n obtained from risk assessment; ω_{ni} is the influence degree of vulnerability of bus n on that of bus i, which is also the maximum value in bus n's corresponding matrix B_k and, finally, R_I is the comprehensive vulnerability value of bus i.

The sorted map of bus low-voltage vulnerability is obtained through the calculation of all the buses. Set the unit time of failure rate as one year. Every time faults of a single line are considered, excluding the case of faults occurring in two or several lines at the same time, namely, calculate the probability of each line occurring fault for one time within a year. The low-voltage risk value of an isolated bus is calculated through a risk assessment process, and then the effect coefficient is calculated considering the mutual influence of bus vulnerability, and, finally, based on formula (6.8), a comprehensive risk index value is obtained.

6.5 EXAMPLE ANALYSIS

A 28 nodes system of a 10 kV distribution network is chosen to carry out the simulation analysis, whose electric circuit is shown in Figure 6.4 [15]. Node 1 is a line outlet with voltage of 10 kV. Figure 6.1 lists the parameters of lines; Figure 6.2 shows the parameters of load

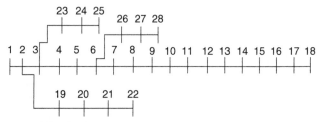

Figure 6.4 Circuit of 28 nodes in a 10 kV distribution network.

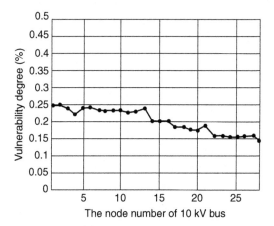

Figure 6.5 Low-voltage vulnerability assessment curve under traditional risk theory.

nodes. The curve of vulnerability indicators achieved through simulation is shown in Figure 6.5 and Figure 6.6, of which Figure 6.5 is the result of bus low voltage vulnerability in the distribution network, evaluated by traditional risk theory, and Figure 6.6 adopts the result of bus low-voltage vulnerability evaluated through fuzzy AHP, mentioned in this chapter (Tables 6.1 and 6.2).

From Figures 6.5 and 6.6, fuzzy AHP gets the similar vulnerability analysis result as the traditional algorithm, except that its indicator value is greater than that of the traditional risk theory, for it takes into account the influence of relevant bus vulnerability that, however, doesn't affect the nature of the risk assessment.

As shown in Figure 6.5, the curve of bus risk indicators achieved through traditional risk theory is relatively flat, and the vulnerability indicator values of node 2, 3, 5, 6 are a bit bigger than those of other nodes, due to their own heavy loads (due to its heavy load, current

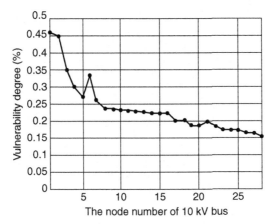

Figure 6.6 Low-voltage vulnerability assessment curve under fuzzy AHP.

flowing through the node is large, and will cause greater voltage loss in the case that line impedance is constant, so low-voltage risk is greater). Generally, bus 1 is a radial distribution network outlet with large risk value. Based on the experience that risk value less than 0.235 is considered as an acceptable risk value, node 1, 2, 3, 5, 6 are buses at high risk, yet the other nodes are relatively much safer.

As shown in Figure 6.6, values of vulnerability indicators of nodes ahead that are obtained through fuzzy AHP in this chapter are relatively higher, and they are yet becoming gentler after node 7. Among them, the risk value of node 2 is the greatest, excepting the initial node, as it is not only heavily loaded, but also related to the other four nodes; the situation of node 3 is similar to node 2, yet there are greater changes in the risk value of node 6 as the fuzzy AHP takes into account the influence between bus vulnerability, meaning that the risk value of node 6 is affected by the vulnerabilities of buses connected to it, respectively, increasing its risk value. Similarly, the vulnerability values of node 5 and 7 are increased due to the influence of node 6 vulnerability. Because in AHP simulation results the risk values of all the buses are higher, lines with risk values above 0.26 are selected as high risk lines of low-voltage, combined with the experience of field personnel, so from Figure 6.6 it can be seen that nodes 1, 2, 3, 4, 5, 6, 7 are lines at high risk. Compared with the results of conventional methods, buses at high risk are increased, and predictions on the

Table 6.1 Parameters of lines

Branch number	Initial node	End node	R (Ω)	X (Ω)	Branch number	Initial node	End node	R (Ω)	X (Ω)	Branch number	Initial node	End node	R (Ω)	X (Ω)
1	1	2	1.8216	0.758	10	10	11	2.752	0.778	19	19	20	1.376	0.389
2	2	3	2.227	0.9475	11	11	12	1.376	0.389	20	20	21	2.752	0.778
3	3	4	1.3662	0.5685	12	12	13	4.128	1.167	21	21	22	4.9536	1.4004
4	4	5	0.918	0.379	13	13	14	4.128	0.8558	22	22	23	3.5776	1.0114
5	5	6	3.6432	1.516	14	14	15	3.0272	0.778	23	23	24	3.0272	0.8558
6	6	7	3.7324	1.137	15	15	16	2.752	1.167	24	24	25	5.504	1.566
7	7	8	1.4573	0.6064	16	16	17	4.128	0.778	25	25	26	2.752	0.778
8	8	9	2.7324	1.137	17	17	18	2.752	0.778	26	26	27	1.376	0.389
9	9	10	3.6432	1.516	18	2	19	3.44	0.9725	27	27	28	1.376	0.389

Table 6.2 Parameters of the node loads

Node number	P (kW)	Q (kVA)	Node number	P (kW)	Q (kVA)	Node number	P (kW)	Q (kVA)	Node number	P (kW)	Q (kVA)	Node number	P (kW)	Q (kVA)
2	1400	90	8	90	50	14	70	40	20	50	30	26	40	20
3	80	50	9	80	50	15	70	40	21	50	30	27	40	20
4	80	60	10	90	50	16	70	40	22	50	30	28	40	20
5	100	60	11	80	50	17	60	30	23	50	30			
6	80	50	12	80	40	18	60	30	24	50	20			
7	90	40	13	90	50	19	70	40	25	60	30			

low-voltage risk values of other nodes also become more accurate. In addition, fuzzy AHP amplifies vulnerability index values, and is more conducive for operating personnel to treat them differently, and to focus on the implementation of security monitoring.

Therefore, fuzzy AHP predicts more accurately the vulnerability degree of low-voltage bus, with increased reliability and accuracy, compared to the traditional algorithm. Dispatchers can take effective measures in advance in order to ensure the safe and economic operation of the distribution network, thus reducing the occurrence of power blackouts.

BIBLIOGRAPHY

[1] Y. Yin, J. Guo, J. Zhao, et al. Preliminary analysis of the North American "8.14" blackouts and lessons should be learned, Power Syst. Technol. 27 (10) (2003) 8–11.

[2] W. Chen, Q. Jiang, Y. Cao, et al. Risk assessment of voltage collapse in power system, Power Syst. Technol. 29 (19) (2005) 6–11.

[3] A. Wang, Y. Luo, G. Tu, P. Liu, Vulnerability assessment scheme for power system transmission networks based on the fault chain theory, IEEE Trans. Power Syst. 26 (1) (2011) 442–450.

[4] Q. Liu, Q. Liu, Q. Huang, J. Liu, Assessment of grid inherent vulnerability considering open circuit fault under potential energy framework, J. Central South Univ. Technol. 17 (6) (2010) 1300–1309.

[5] X. He, J. Peng, Y. Gong, Study on the misunderstandings of reducing line losses of distribution network, Power Syst. Prot. Control 38 (1) (2010) 96–99.

[6] N. Wang, W. Chen, L. Luo, Low voltage security warning based on risks of power system, East China Electr. Power 36 (3) (2008) 66–69.

[7] D.P. Bernardon, M. Sperandio, V.J. Garcia, et al. AHP decision-making algorithm to allocate remotely controlled switches in distribution networks, IEEE Trans. Power Deliver. 26 (3) (2011) 1884–1892.

[8] W. Pedrycz, M. Song, Analytic hierarchy process (AHP) in group decision making and its optimization with an allocation of information granularity, IEEE Trans. Fuzzy Syst. 19 (3) (2011) 527–539.

[9] D.H. Spatti, I.N. da Silva, W.F. Usida, R.A. Flauzino, Real-time voltage regulation in power distribution system using fuzzy control, IEEE Trans. Power Deliver. 25 (2) (2010) 1112–1123.

[10] D.H. Spatti, I.N. da Silva, W.F. Usida, R.A. Flauzino, Fuzzy control system for voltage regulation in power transformers, IEEE Lat. Am. Trans. 8 (1) (2010) 51–57.

[11] Q. Li, L. Zhou, F. Zhang, et al. Comprehensive assessment on forewarning grade of VST based on fuzzy theory and analytical hierarchy process in power system, Power Syst. Technol. 32 (4) (2008) 40–45.

[12] B. Liu, T. Zhu, J. Yu, Multi-level fuzzy comprehensive evaluation on forewarning grade of VST in power system, Power Syst. Technol. 29 (24) (2005) 31–36.

[13] I. Kamwa, A.K. Pradhan, G. Joos, S.R. Samantaray, Fuzzy partitioning of a real power system for dynamic vulnerability assessment, IEEE Trans. Power Syst. 24 (3) (2009) 1356–1365.

[14] E. Bompard, R. Napoli, F. Xue, Extended topological approach for the assessment of structural vulnerability in transmission networks, Gener. Transm. Distrib. 4 (6) (2010) 716–724.

[15] Z. Wang, J. Cai, Power flow calculation of radial distribution network based on threaded binary tree, High Voltage Eng. 32 (6) (2006) 113–118.

CHAPTER 7

Direction-Coordinating Based Ant Colony Algorithm and its Application in Distribution Network Reconfiguration

Contents

7.1 INTRODUCTION

As one of the infrastructures facing users in the power system directly, distribution network employs the functions of distributing power energy to users. Because of its complex network structure, and the large line losses due to its long lines, to reduce energy losses during the transmission process – without lessening its safety and quality – is also an important aspect to be considered when designing the structure of a distribution network.

A closed structure is used when designing the distribution network, yet an open structure is adopted during its operation; this not

X. Meng and Z. Pian: Intelligent Coordinated Control of Complex Uncertain Systems for Power Distribution Network Reliability. http://dx.doi.org/10.1016/B978-0-12-849896-5.00007-6

only reduces fault coverage effectively, but also makes the operation mode of the distribution network more flexible. In the network structure of the distribution network, there are many segments and interconnection switches, whose opening and closing states are changed to choose a different power supply during the operating process of the distribution network, causing loads of lines to transfer at the same time. This process is called the reconfiguration of the distribution network. Distribution network reconfiguration is aimed at removing line overloads, reducing power losses on the network, as well as improving transmission quality, and so on.

Since the distribution network has characteristics such as complex structure, long lines, and big losses, the distribution network reconfiguration problem can be seen as a complex combinatorial optimization problem bound with multiple constraints. To solve this problem, both domestic and overseas scholars have made extensive researches. Methods able to solve this problem currently include the mathematical optimization method, the heuristic approach, and the artificial intelligence algorithm. The mathematical optimization method commonly uses existing mathematical optimization principles to solve the problem. Although this method can get the global optimal solution, the "combinatorial explosion" problem may appear when solving this issue because of the complexity of the distribution network reconfiguration, as the scale of the problem increases. Besides, it takes a long time to calculate a result, so there are certain difficulties in the practical application. The heuristic approach mainly includes the optimal flow pattern method [1] and the branch exchange method [2] that, however, is difficult to guarantee a global optimum in solving the problem of distribution network reconfiguration. In recent years, the artificial intelligence algorithm has developed a new effective method of solving problems in distribution network reconfiguration [3] that currently includes the genetic algorithm [4–6], the simulated annealing algorithm [7], the *tabu* search algorithm [8,9], and particle swarm optimization [10], and so on.

The ant colony algorithm is also a kind of artificial intelligence algorithm, and there are scholars applying it to solve distribution network reconfiguration problems, but, compared to other artificial intelligence algorithms, ant colony algorithm's application in distribution network reconfiguration is relatively less encountered. Improved ant colony algorithm based on directional pheromone is used to solve distribution network reconfiguration problems in this chapter, and it is also adopted in order to analyze the model of distribution network reconfiguration. Aimed at reducing distribution network losses, the constraints in reconstruction problems are substituted, and simulation experiments are undertaken in order to prove that the improved ant colony algorithm based on directional pheromone is able to achieve optimum results when solving distribution network reconfiguration problems.

7.2 IMPROVED ANT COLONY ALGORITHM BASED ON DIRECTIONAL PHEROMONE

7.2.1 Improved Strategies of Pheromones

In the ant colony algorithm (ACS algorithm), since ants only update the pheromones on the optimal route, accelerating the convergence speed of algorithm to some extent – which, however, cannot achieve global updating since it only strengthens the information feedback from the optimal path. As such, it is very easy for the phenomenon of stagnation to happen. In order to avoid the algorithm causing stagnation and speeding up its convergence rate at the same time, this section presents a method combining regional update and global update of pheromones. During the process of updating the ordinary pheromones through regional information, a new pheromone is defined in the algorithm – the directional pheromone that will be updated based on global information. After combining the two pheromones, pheromone values between city nodes of shorter distances will also be strengthened, in addition to the pheromones on the optimal path; this not only strengthens the information feedback from the optimal path and accelerates the algorithm's convergence, but also inhibits stagnation in the algorithm, to a certain extent.

In this algorithm, traversing ants will not only leave directional pheromones on each city, but also release common pheromones along the path. Superimposition of ordinary pheromones is used to indicate the ants' access frequency to this path, and directional pheromones on each city node give a global directional alternative that not only speeds up the convergence rate, but also prevents effectively the algorithm falling into regional optimization during its operating process.

1. Updating principles of ordinary pheromones

   During traversal, the ant will update pheromones on the passed route upon each access to the city, and the updating principles of pheromones are shown in expression (7.1) and expression (7.2):

$$\tau_{ij}(t+n) = (1-\rho)\tau_{ij}(t) + \Delta\tau_{ij}(t) \tag{7.1}$$

$$\Delta\tau_{ij}(t) = \begin{cases} \dfrac{Q}{d_{ij}} \text{ ants have passed}(i,j) \\ \\ 0 \text{ ants haven't passed}(i,j) \end{cases} \tag{7.2}$$

   In the expression, Q is a constant, the concentration of ordinary pheromones; ρ is the volatile rate of pheromones, ranged from 0 to 1; d_{ij} is the distance between city i and j.

2. Updating the principles of directional pheromones

   After traversal, directional pheromones on the visited city will be updated. Directional pheromones are released on city nodes, represented by vectors in detail. On city i, $\vec{\tau}_{ij}$ stands for the directional pheromone value from city i to city j.

   Different from the principles of updating ordinary pheromones, directional pheromones in the algorithm don't superimpose or evaporate, and their updating rules are: after ant k has completed a traversal, it will update directional pheromone values on each city node, according to the order of the visited cities in this traversal. Assume city j as the next node to city i in the current traversal path, so before updating the directional pheromone value from city i to city j there is $\vec{\tau}_{ij}$. Update the pheromone value on city node i according to expression (7.3); this means that ants will compare the new directional pheromone value with that of the previous city, after completing a traversal. If the former is larger, it will update, otherwise it keeps the current directional pheromone value.

$$
\vec{\tau}_{ij}(t+n) = \begin{cases} \dfrac{Q'}{L_k} & \dfrac{Q'}{L_k} > \vec{\tau}_{ij}(t) \\[4mm] \vec{\tau}_{ij}(t) & \dfrac{Q'}{L_k} \le \vec{\tau}_{ij}(t) \end{cases} \tag{7.3}
$$

In the expression, Q' is the concentration of directional pheromones, which is a constant; and L_k is the length of the path traversed by ant k.

7.2.2 Improved Selection Strategies

After adding directional pheromones, the probability formula of selection also needs appropriate improvements. In addition, a new exploration rate factor is proposed during the path selection process in order to expand the initial search range of the algorithm, reducing the possibility of local optimum.

1. Improvement of selection probability formula

   After adding directional pheromones, ants must comprehensively consider directional pheromone and ordinary pheromone, when selecting a path, so directional pheromone information needs to be accurately added to the probability formula of the ant's selection. Due to different pheromone updating rules of ordinary and directional pheromones, ordinary pheromones reflect the access probability to local routes, whereas directional ones show global information. So, although both ordinary and directional pheromones play the roles of positive feedback on optimal information, the two should have different heuristic factors in the probability formula of the ant's selection of routes. Therefore, the probability formula of the ant's selection of paths is improved, as shown in expression (7.4), based on which ants calculate the probability of visiting each city during the process of choosing paths.

$$
p_{ij}^k(t) = \begin{cases} \dfrac{\tau_{ij}^{\alpha}(t)\eta_{ik}^{\beta}(t)\vec{\tau}_{ij}^{\gamma}(t)}{\displaystyle\sum_{s \subset allowed_k} \tau_{is}^{\alpha}(t)\eta_{is}^{\beta}(t)\vec{\tau}_{is}^{\gamma}(t)} & j \in allowed_k \\[6mm] 0 & j \notin allowed_k \end{cases} \tag{7.4}
$$

In the expression, *allowed$_k$* is the allowed threshold value; α reflects the effect of information accumulated by the ant in its traversal process on its movement; β reflects the degree in which heuristic information is valued in the ant's selection of paths during its movement; γ reflects the influence of directional pheromones on the ant's selection of paths.

In the ant's traversal process, the influence of each factor on the ant's selection of paths may be tuned by appropriately adjusting the values of α, β, and γ.

2. Exploration rate factor

To avoid the algorithm falling into local optimum too early, an exploration rate value is added during the selection of paths in the improved algorithm, whose value can be decreased progressively, according to expression (7.5)

$$\varepsilon(t+n) = \varepsilon(t) - c \qquad (7.5)$$

In the expression, c is a constant.

Ants will update the value of ε after completing each traversal.

When the ant chooses a path in a traversal process, it will produce a random number ranged between $0 \sim n$ (the number of cities). The ant will compare the random number with ε and, if it is smaller than ε, it will choose to generate a city randomly, otherwise it will select based on expression (7.4).

Since ε is a progressively decreasing value, in the initial stage during the ant's selection of paths, the chance to select a random number is bigger, and increases greatly the algorithm's searching range, reducing the probability of falling into local optimum in the algorithm's searching process. In addition, with the increasing of the traversal's times, ε will eventually reduce to 0. Pheromones on the optimal route will increase after the ants' selections through expression (7.4), and will converge to an optimal value finally while the ants are selecting paths.

7.2.3 Realization of the Algorithm

The steps to realize the programming of improved ant colony algorithm (its process is shown in Figure 7.1) are:

1. Step 1: Initialize the environmental information. Set the directional pheromone value on each city node as 0, the pheromone

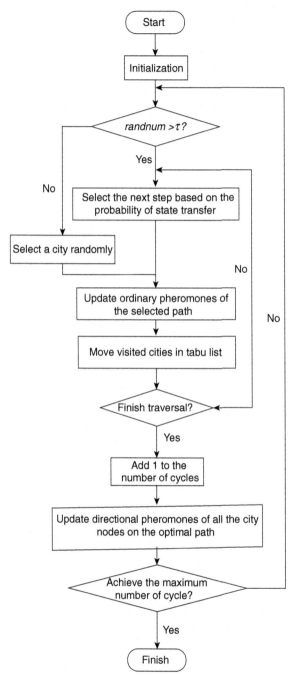

Figure 7.1 Process of improved ant colony algorithm.

value on each path $\tau_{ij}(t) = const$, of which *const* stands for a constant. Make the initialized value $\varepsilon = N$, of which N is a fixed integral value. Put m ants on n cities at the initial time $\Delta\tau_{ij}(0) = 0$.

2. Step 2: Compare the size of random number (*randnum*) and exploration rate (ε). A random number ranged between $0\sim n$ is produced, and if it is greater than the exploration rate, the individual ant will choose city j, marching forward based on the probability calculated through expression (7.4), otherwise it will choose to produce a city randomly, and meanwhile the value of exploration rate ε is updated through expression (7.5).

3. Step 3: Modify the *tabu* lists. After choosing a city, the ant is moved to a new city, and the city is moved to its *tabu* list. Update the pheromone value of the chosen path based on expression (7.1) and (7.2).

4. Step 4: Judge whether it has completed the traversal. Judge whether the cities in set C have finished traversal, and if they haven't, skip to step 2, otherwise execute step 5.

5. Step 5: Add 1 to the number of cycles, and update the pheromone value on each city node through expression (7.3).

6. Step 6: If finishing conditions are satisfied, the maximum number of cycles N_{max} is reached, so finish cycling and output the result, otherwise empty the *tabu* list and skip to step 2.

7.2.4 Performance Analysis of Algorithm's Parameters

1. The impact of γ's valuing on algorithm

   From expression (7.4), it can be seen that the value of γ determines the influence of newly added directional pheromone's value on the ant's selection of routes, and the size of this parameter greatly affects the convergence rate of algorithm, after directional pheromones are added. As such, simulation analysis about the influence of the size of γ on the algorithm is undertaken through experiments.

   In the improved algorithm, since the directional pheromones are not superimposed or volatile, yet ordinary pheromones are constantly superimposed, the volatile coefficient ρ of ordinary

Table 7-1 Impact of the valuing of γ on the algorithm

γ	Optimal value	Worst value	Mean value
0	456.36	462.72	459.48
5	441.33	460.1	452.83
10	441.48	450.93	446.51
15	431.98	458.37	444.89
20	433.2	457.74	444.28
25	431.98	451.84	442.51
30	438.85	455.17	444.55
50	433.2	460.1	457.32

pheromone is set to a bigger value in the experiment, in order to balance the impacts of these two kinds of pheromones on the ant's selection of paths. The size of ordinary pheromone is connected with the distance between two cities, while the size of directional pheromone is affected by the overall traversal path, so when setting the values of Q and Q' that stand for the pheromone concentration, the value of Q' is set as n times of Q's value.

Select Eil51 as the experimental subject, and make $\alpha = 1$, $\beta = 5$, $\rho = 0.9$, $\varepsilon = 100$, $Q = 10$, $Q' = 100$, $N_{max} = 1000$, the number of ants $m = 51$. Simulation analysis is carried out considering different values of γ, and the results are shown in Table 7.1. The optimal value, the worst value, and the mean value in Table 7.1 refer to the shortest, the longest, and average values of the shortest path, respectively, achieved from 30 iterations of the simulation experiments.

In the case of Q' = 100, when $\gamma = 0$, directional pheromones have no effect on the algorithm's solution; when $\gamma = 5$, the accuracy of the algorithm's solution is obviously increased; when $\gamma = 10$, the mean value of the algorithm's solution is improved by 2%, compared to the case when $\gamma = 0$; as the value of γ further increases, the accuracy of the algorithm's solution also improves gradually, and when $\gamma > 15$, the accuracy of the mean value obtained from the algorithm has been improved by 4%, compared to the case when $\gamma = 0$; when $\gamma = 25$, the algorithm converges to 431.98, with the error of 1%, reaching the optimal accuracy.

As such, it can be noted that as the value of γ increases, the directional pheromone's influence on the algorithm's solution also increases, along with its accuracy. When $\gamma > 15$, the solving performance of the algorithm tends toward stability, but after $\gamma > 30$, regional optimal begins to occur in the algorithm's solution process, with a greater difference between the optimal solution and the worst solution, so it is very unstable. Therefore, of all these 30 experimental results, the solving performance of the improved algorithm is better when $\gamma = 25$.

2. Impact of the valuing of ε on the algorithm

The valuing of ε determines the size of the searching range at the initial stage of the improved algorithm, and affects its convergence rate in the meantime. If its value is too large, the initial searching range of the algorithm is wide, reducing the convergence speed at the same time; if its value is too small, the algorithm can easily fall into the regional optimum at the initial stage, a fact that is unfavorable for the algorithm solution. So, analyzing valuing of ε is needed to explore its relationship with the number of cities.

Take Eil51 as example again, and make $\alpha = 1$, $\beta = 5$, $\rho = 0.9$, $Q = 10$, $Q' = 100$, $\gamma = 25$, the number of ants $m = 51$ and subtract 1 from the value of ε value with every five traversals. Simulation experiments are carried out to discuss the relationship between the initial value of ε and the number of cities; and, in this experiment, the cities' number n is 51, with the experimental results shown in Table 7.2. The results presented in Table 7.2 are the maximum, minimum, and mean value of the shortest path achieved from 30 experiments, and the number of iterations stands for the mean value of iterations when the algorithm tends to converge in 30 experiments.

Table 7.2 Impact of the valuing of ε on algorithm

ε	25(0.5n)	51(n)	75(1.5n)	100(2n)
Optimal value	433.35	438.13	429.74	431.98
Worst value	448.28	458.05	449.82	448.31
Mean value	442.09	445.12	437.20	441.31
Number of iterations	793	830	1025	1162

From Table 7.2, it can be seen that when $\varepsilon = 25(0.5n)$, the number of iterations of convergence is 793, with a faster convergence rate, and the optimal value achieved is 433.35, with an error of 1.7%, and better accuracy; when $\varepsilon = 51(n)$, the number of iterations is 830, with the optimal value of 438.13 and an error of 2.8%, whose accuracy is reduced, and iterations are increased, compared to the case of $\varepsilon = 25$; when $\varepsilon = 75(1.5n)$, the optimal value of the algorithm's solution is 429.74, with an error of 0.9%, and better accuracy; however, when $\varepsilon = 100(2n)$, the optimal value is 431.98, with an error of 1.4%, and worse accuracy, compared to the case of $\varepsilon = 75$, and furthermore, when the algorithm tends to converge, the iterations will obviously increase. So, when ε is smaller or equal to the number of cities, the convergence rate of the improved algorithm is greater, but its result of the shortest path achieved from the algorithm is also correspondingly unstable; whereas, when it is larger than the number of cities, the algorithm's convergence rate will fall a bit, however, the result of the corresponding shortest path is optimal. But when ε is too large, the algorithm's performance of search optimum will also drop, relatively. Analyzing the above experiments comprehensively, it can be said that when the value of ε is 1.5 times that of the number of cities, the algorithm's solving performance is better.

7.2.5 Check the Simulation and Operation Results

1. The impact of improved strategies on convergence rate

In order to test the convergence rate of the algorithm proposed, we take Eil51 as experimental subject, and compare the searching results of the improved algorithm added with directional pheromones, with that of the algorithm with nondirectional pheromones, under the condition that the values of ε are the same; the experimental results are shown in Figure 7.2.

Since the ant traversal process is largely random at the initial stage, the value of the path length achieved from the initial stage is greater. However, as iterations increase, the algorithm can quickly converge. From Figure 7.2, it can be seen that the algorithm begins to converge after 400 iterations, and converges eventually after about 1500 iterations, when none of the directional

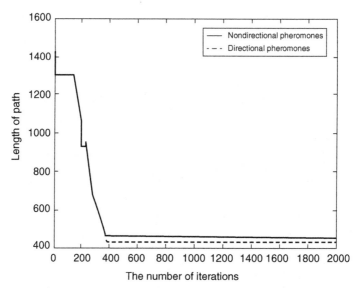

Figure 7.2 Comparison between the algorithm added with directional pheromones and the one with nondirectional pheromones.

pheromones are added; in the improved algorithm, added with directional pheromones, the algorithm is able to converge to 429.74 after about 1000 iterations. Therefore, compared with the algorithm with no directional pheromones added, the convergence rate of the algorithm after adding directional pheromones improves significantly, so does the accuracy of the algorithm.

For the issue of convergence rate, simulation experiments of the improved algorithm and ACS algorithm are compared, and results are shown in Table 7.3. The minimum, maximum, and mean values

Table 7.3 Comparison between the convergence rate of improved algorithm and that of ACS algorithm

TSP problem		Oliver 30	Eil51
ACS algorithm	Minimal value	636	975
	Maximal value	1851	1737
	Mean value	1267	1421
	Average solution	429.83	441.64
Improved ant colony algorithm	Minimal value	191	450
	Maximal value	531	1703
	Mean value	306	1078
	Average solution	424.64	437.20

in Table 7.3 refer to the iterations when the algorithm tends to converge in 30 experiments, and the average solution refers to the average value of the shortest paths achieved from 30 experiments.

From Table 7.3, it can be seen that in an Oliver30 problem the mean value of iterations in the improved algorithm is 306, with the average value achieved of 424.64, about a quarter of the ACS algorithm's, with the error reduced by 1.2% at the same time; in the Eil51 issue, the average value of iterations in the improved algorithm is 1078, with the mean value achieved of 437.20, about three quarters of the ACS algorithm's, with the error reduced by 1.0%. Therefore, the improved algorithm can get better solutions, as well as obviously faster convergence rates, that benefited from the directional pheromones added. In an improved algorithm, ordinary pheromones update according to local information, whereas directional pheromones update according to integrated overall information. After the combination of the two, pheromone values on the paths of smaller weights, as well as the shortest paths, are both strengthened, fact that encourage ants to converge toward the shortest path, improving the convergence rate of the algorithm during the process of searching optimum.

2. Performance analysis of improved algorithm

Improved ant colony algorithm and other ant colony algorithms are applied, respectively, into some typical TSP problems in TSPLIB in order to carry out simulation experiments. Set the algorithm parameter $m = n$ (the number of ants is equal to that of cities), and make $\alpha = 1$, $\beta = 5$, $\rho = 0.9$, $\varepsilon = 1.5n$, $Q = 10$, $Q' = 100$, and the detailed comparison results of algorithm simulation experiments are shown in Table 7.4.

For various TSP problems, the improved algorithm is able to find a shorter path than the ACS algorithm, demonstrating a good overall searching ability. From Table 7.4 we can see that, in an Oliver30 problem, the shortest path length achieved from the improved ant colony algorithm is 423.9, with the error dropping by 0.4% from the ACS algorithm's result of 425.5, and, at the same time, its iterations are only one third of the ACS algorithm's; besides, in the Eil51 problem, the error of optimal value achieved from the improved algorithm drops by 2%, and its average

Table 7.4 Table of detailed comparisons between improved algorithm and each ant colony algorithm

TSP example	Name	Length of the shortest path	Error (%)	Length of the worst path	Average length of paths	Average iterations
Oliver30	ACS algorithm	425.52	0.6	446.87	429.83	1267
	Improved algorithm	423.91	0.2	425.82	424.64	497
Eil51	ACS algorithm	438.74	2.9	455.17	441.64	1422
	Improved algorithm	429.74	0.9	449.82	437.20	1087
St70	ACS algorithm	687.46	1.8	700.95	697.37	1282
	Improved algorithm	682.31	1.0	699.12	689.92	932
Eil76	ACS algorithm	555.61	3.3	559.55	557.18	1272
	Improved algorithm	544.43	1.2	557.43	550.21	970

iterations fall by one quarter, compared to the ACS problem. As the number of cities in a TSP problem increases, the error of the improved algorithm also gets bigger, but its convergence rate is obviously enhanced, and its results are more accurate than that of the ACS algorithm. In conclusion, the improved algorithm is more accurate, with a better overall searching capacity.

7.3 APPLICATION OF DIRECTION-COORDINATING ANT COLONY ALGORITHM IN DISTRIBUTION NETWORK RECONFIGURATION

7.3.1 Distribution Network Reconfiguration Issue

Distribution network reconfiguration refers to the situation when, under the condition that all the constraints to ensure the safe operation of the power grid are satisfied – such as requirements for voltage drop, line heat capacity, etc. – power supply and lines are chosen by changing the opening and closing states of each segment switch and interconnection switches, aimed at forming eventually a radial network through this choice. This is so that it may make a particular index of the distribution network, such as the line losses, voltage capacity, load balancing, and so on, achieve the optimal state.

The distribution network reconfiguration problem studied in this chapter selects distribution network losses as the objective function, in order to apply the directional pheromone-based ant colony algorithm in distribution a network reconfiguration problem, for the purpose of reducing distribution network losses.

The distribution network reconfiguration problem is rendered abstract. The distribution transformer, feeder sections, and loads of the network are considered as nodes, and circuits as edges in the figure. As such, the distribution network is represented as an undirected connected graph $G' = (V, E)$, where V represents a set of individual nodes, and E represents the set of edges in the network. Thus, the optimal solution that needs to be solved in the distribution network reconfiguration problem is rendered abstract, as searching a minimal spanning tree G that meets the minimal conditions of network losses under the premise that some constraints are satisfied, and the G generated is a directed graph.

The network losses of the distribution network include copper and iron losses of the transformers' own consumption, and losses of line conductors, and so on, yet only line conductor losses can be changed through distribution network reconfiguration, so this chapter chooses loss minimum as an objective function, shown in expression (7.6):

$$\min f_{loss} = \sum_{i=1}^{N_b} k_i R_i \left| I_i \right|^2 = \sum_{i=1}^{N_b} k_i R_i \frac{P_i^2 + Q_i^2}{U_i^2} \tag{7.6}$$

In the expression, N_b is the number of branches in the distribution network; R_i is the resistance on branch i; P_i and Q_i are the active power and reactive power on branch i, respectively; U_i is the node voltage at the end of branch i; k_i is the discrete variable of $0 - 1$ used to represent the opening and closing state of switch i, where 0 stands for open, 1 stands for closed; and I_i is the current on branch i.

When distribution network reconfiguration is carried out, the following constraints need to be satisfied:

1. Network topology constraint. After the distribution network is reconfigured, the network formed must be radial.
2. Power supply constraint. The reconfigured distribution network must satisfy the requirements of line loads, without any independent nodes in the distribution network at the same time.
3. Inequality constraint. It includes constraint of node voltage [equation (7.7)], overload constraint of branch [equations (7.8) and (7.9)] and overload constraint of transformers [equation (7.10)], and so on.

$$U_{i\min} \leq U_i \leq U_{i\max} \tag{7.7}$$

$$S_i \leq S_{i\max} \tag{7.8}$$

$$I_i \leq I_{i\max} \tag{7.9}$$

$$S_t \leq S_{t\max} \tag{7.10}$$

In the expression, $U_{i\max}$ and $U_{i\min}$ are the upper and lower limit values of voltages allowed on node i; S_i and $S_{i\max}$ are the value calculated from power flowing through each branch, and its maximal allowed value; I_i and I_{imax} are the current flowing through branch i and its maximal allowed value; S_t and S_{tmax} are the output power of the transformer and its maximal allowed value.

7.3.2 Construct the Minimal Spanning Tree by Ant Colony Algorithm

The result solved by the distribution network reconfiguration problem is a radial network meeting constraints, and the process of building the radial network is similar to that of constructing a minimum spanning tree. Set $S_k(t)$ to denote the set of nodes when ant k is connected to the tree at time t, corresponding to the *tabu* list $tabu_k$ in the ant colony algorithm; $W_k(t)$ represents the set of nodes when ant k isn't connected to the tree, a set of all the candidate nodes; $E_k(t)$ is the set of all the feasible paths between the sets at time t, also a set of all feasible solutions under current states; $P_k(t)$ is the possibility value of state transfer on each path at time t; $A_k(t)$ is a new set of optional edges added in set $E_k(t)$ at time t; $s - w$ stands for an edge from node s to node w, which are vertexes of the edge, respectively.

Steps of constructing a minimal spanning tree by ant colony algorithm are as follows:

Step 1: Set $t = 0$, and ant k sets off from the starting point, so $S_k(0) = \{s_0\}$.

Step 2: Ant k chooses the edge $j(s - w)$ from the set $E_k(t)$ based on $P_k(t)$.

Step 3: Check if there is a line connected to node w in the set $E_k(t)$, and if there is, disconnect j and return to step 2, otherwise execute step 4.

Step 4: Update the set of nodes and $W_k(t)$, and move node w from the set $W_k(t)$ to $S_k(t)$, so $W_k(t+1) = W_k(t) - \{w\}, S_k(t+1) = S_k(t) + \{w\}$.

Step 5: Check whether set $W_k(t)$ is empty, and if it is, all the load nodes have been connected to the spanning tree and the whole process is finished, otherwise execute step 6.

Step 6: Update set $E_k(t)$ and remove the edge j from set $E_k(t)$, and add the new set of alternative edges $A_k(t)$ into set $E_k(t)$, so $E_k(t+1) = E_k(t) + A_k(t) - \{j\}$.

7.3.3 Realize the Application of Direction-Coordinating Based Ant Colony Algorithm (Dual Population Ant Colony Optimization, DPACO) in Distribution Network Reconfiguration

According to the requirements of network topology, and power supply constraints in distribution network reconfiguration, the reconstructed distribution network must include all of the nodes in the distribution network, and its solutions are required to be radial, with no return circuits. As such, the ant colony algorithm can be used to traverse the network, and can also be used to construct a minimal spanning tree, and, at the same time, in order to accelerate the implementation efficiency of the algorithm, the aforementioned ant colony algorithm based on directional pheromones is applied to the distribution network reconstruction problems.

1. The pheromone update

    During the updating of the pheromones, the improved algorithm contains two kinds of pheromones, one that is of the ordinary type, and the other that is the directional pheromones. The updating method of this kind of pheromones is adjusted appropriately, considering the problems in distribution network reconfiguration.

    a. The update of ordinary pheromone. In the distribution network, at the initial time, for pheromones on each path in the initialized network, set $\tau_{ij} = d, d$ is a constant. After completing a traversal, the ant will get a minimal spanning tree to generate a radial network. Data achieved from the network are processed to get the loss value f_{lossi} of the path obtained in the iteration through power flow calculation. Different from TSP problems, each branch is different from each other in the distribution network, so, when choosing a path, the resistance value R_{ij} on each branch must be taken into account.

        After finishing a traversal, the ant k will update pheromones on each path in the network formed, currently based on expression (7.11) and (7.12).

$$\tau_{ij}(t+n) = (1-\rho)\tau_{ij}(t) + \Delta\tau_{ij}(t) \qquad (7.11)$$

$$\Delta\tau_{ij}(t) = \begin{cases} \dfrac{QR_{ij}}{f_{\text{lossk}}} & \text{the ant passes } (i,j) \\[2mm] 0 & \text{the ant doesn't pass } (i,j) \end{cases} \tag{7.12}$$

In the expression, f_{lossk} is the loss value of the radial network formed by ant k after the traversal; R_{ij} is the resistance value on lines between nodes i and j; ρ is the volatility of pheromones.

By updating the ordinary pheromones, pheromones on each branch can be controlled to record the paths contained in the feasible solutions, thus speeding up the algorithm's solution.

b. The directional pheromone update. After completing a traversal, ants will update the directional pheromones on each branch in the radial network formed. Because directional pheromone mainly plays a guiding role that is not volatile, only the path with the minimal value of network losses is chosen in order to get updated, from all the paths formed by the ants in this time of iteration during the updating of directional pheromones in the network.

The updating rule of directional pheromones follows the expression (7.13).

$$\vec{\tau}_{ij}(t+n) = \begin{cases} \dfrac{Q'R_{ij}}{f_{\text{bestN}_c}} & \dfrac{Q'R_{ij}}{f_{\text{bestN}_c}} > \vec{\tau}_{ij}(t) \\[4mm] \vec{\tau}_{ij}(t) & \dfrac{Q'R_{ij}}{f_{\text{bestN}_c}} \leq \vec{\tau}_{ij}(t) \end{cases} \tag{7.13}$$

In the expression, $f_{\text{bestN}c}$ is the minimal loss value of the radial networks formed by all of the ants; R_{ij} is the resistance on branch (i,j); $\vec{\tau}_{ij}(t)$ is the value of directional pheromones on the current branch, before updating.

In this iteration, the loss values of the networks formed by each ant are calculated after all of the ants finish the iteration, and pheromones of each branch on the scheme of the minimal network loss value are updated. During the updating process, the pheromone value on the branch to be updated currently is compared with the value that needs to be updated, and the directional pheromones on the branch with a greater value are chosen to get updated.

2. Strategies to select a branch

   During the traversal process, ants are able to calculate each probability of alternative paths currently, according to expression (7.14).

$$p_{ij}^k(t) = \begin{cases} \dfrac{\tau_{ij}^\alpha(t)\eta_{ik}^\beta(t)\vec{\tau}_{ij}^\gamma(t)}{\displaystyle\sum_{s \subset allowed_k} \tau_{is}^\alpha(t)\eta_{is}^\beta(t)\vec{\tau}_{is}^\gamma(t)} & (i,j) \in E_k(t) \\[4mm] 0 & (i,j) \notin E_k(t) \end{cases} \tag{7.14}$$

In the expression, $allowed_k$ is the allowed threshold value; τ_{ij} is the ordinary pheromone; $\vec{\tau}_{ij}$ is the directional pheromone; η_{ij} is the heuristic factor, whose value is made to be related with the resistance in the problem of distribution network reconfiguration, and the size of its value is determined by expression (7.15).

$$\eta_{ij} = \frac{1}{R_{ij}} \tag{7.15}$$

When ant k is choosing a branch, first of all, a random number rangeing from $0 \sim n$ (the number of nodes) is generated and compared with ε. If ε is smaller, an edge from set $E_k(t)$ is randomly chosen, otherwise the probability value p_{ij}^k of each route in the set is calculated to choose the branch with the biggest value of p_{ij}^k.

3. Realization of DPACO in distribution network reconfiguration

   To realize the application of improved ant colony algorithm based on directional pheromones, in problems of distribution network reconfiguration (the process is shown in Figure 7.3), the steps are as follows:

   Step 1: Read various data values of the distribution network, and regard them as initial statistics of the algorithm.

   Step 2: Initialize the environmental information in the problem of distribution network reconfiguration. Assume the value of the directional pheromone on each node as 0, the pheromone

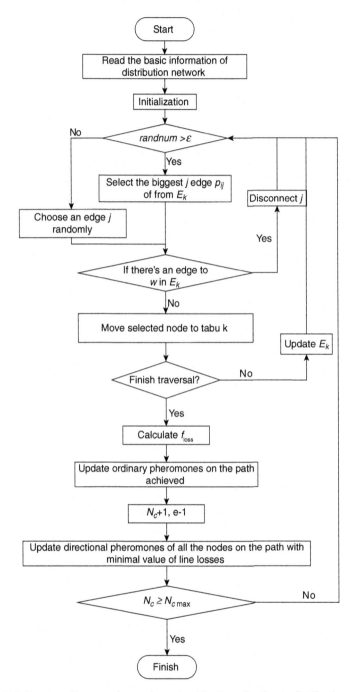

Figure 7.3 Process of improved ant colony algorithm's application in distribution network reconfiguration.

value on each branch as $\tau_{ij}(t) = const$, where *const* is a relatively small constant; set the initial value of ε as $\varepsilon = N$, where N is a fixed integer value, N_{cmax} is the maximum iterations, $\Delta\tau_{ij}(0) = 0$ at the initial time. The *tabu* list $tabu_k$ of each ant k in the ant colony is emptied, and m ants are put randomly on n nodes of the distribution network, with the nodes where each ant is currently found added into the *tabu* list.

Step 3: Compare the size of the random number and exploration rate. A random number ranging between $0 \sim n$ is generated, and if it is greater than the exploration rate, the individual ant will choose edge j from the set $E_k(t)$ of alternative edges, based on the probability calculated through expression (7.14) (assuming the change as $s - w$), otherwise an edge is randomly chosen from the set $E_k(t)$ of alternative edges.

Step 4: Judge whether there is an edge connected to node w in set $E_k(t)$, and if there is, disconnect j and return to step 3, otherwise move on to step 5.

Step 5: Modify the *tabu* list. After an edge is chosen, acting according to it, the ant will move to a new node and put the new node into its *tabu* list.

Step 6: Judge whether it finishes the traversal. Determine whether the set $W_k(t)$ is empty, that is, whether the nodes in the network finish traversal, and if they do, execute step 7, otherwise jump to step 8.

Step 7: Update the set $E_k(t)$ and remove the edge j from set $E_k(t)$, adding the new set $A_k(t)$ of alternative edges into the set $E_k(t)$, that is $E_k(t+1) = E_k(t) + A_k(t) - \{j\}$, and jump to step 3.

Step 8: Add 1 to the number of cycles, and update the value of exploration rate ε according to equation (4.5) and calculate each network loss value; update the values of ordinary pheromones on each node through expression (7.11) and (7.12), in order to find the minimal value of each network losses, and update the directional pheromone values by equation (7.13).

Step 9: If finishing conditions are satisfied, the maximum number of cycles N_{cmax} is achieved, so finish cycling and

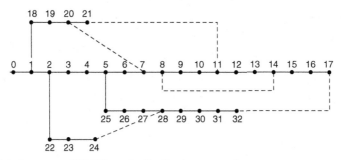

Figure 7.4 Structure of IEEE33-node distribution network.

output calculation results, otherwise empty the *tabu* list and skip to step 3.

4. Sample analysis

An IEEE33-node distribution network is adopted as an example to undertake simulation experiments on the algorithm, whose node structure is shown in Figure 7.4 before the distribution network is reconstructed. Each parameter in the example is: 12.66 kV rated voltage, 33 nodes, 37 branches, 5 interconnection switches, and total loads of 3715 kW + $j2300$ kvar with reference power of 10 MVA.

Assume each parameter as: the number of ants $m = 30$, the initial pheromones on each path $\tau(0) = 0.2$, $\alpha = 1$, $\beta = 5$, $\gamma = 25$, $\rho = 0.6$, $\varepsilon = 50$, $Q = 1$, $Q' = 10$, based on which simulation experiments are carried out with results shown in Figure 7.5.

In Table 7.5, the loss value of the network reconstructed by the improved algorithm decreases visibly by 30% compared to how it was before reconfiguration, improving the performance of the distribution network greatly. Meanwhile, the network loss value during reconstruction by adopting the improved ant colony algorithm is also smaller than that achieved by adopting the basic ant colony algorithm. As the iterations of ants increase, the convergences when the basic and the improved ant colony algorithm calculating network loss value are shown in Figure 7.5. It can be seen that the improved algorithm can be converged after about 35 iterations; this not only takes a shorter time, but also has a smaller network loss value upon convergence than the basic

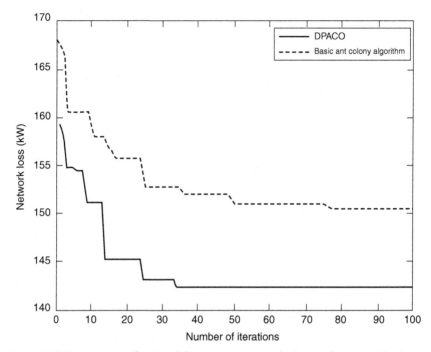

Figure 7.5 Comparison of network loss convergences between dpaco and basic ant colony algorithm.

ant colony algorithm's solution of reconfiguration problem. Experiment shows that using DPACO to reconstruct a distribution network can effectively reduce network loss value with a bigger convergence rate in the circumstances in order to ensure the quality of electricity power.

Table 7.5 Basic ant colony algorithm and calculation results of DPACO simulation experiment

Project	Before configuration	Basic ant colony algorithm	DPACO simulation experiments
Set of opening switches	7–20	6–7	6–7
	8–14	8–9	8–9
	11–21	13–14	13–14
	17–32	31–32	31–32
	24–28	7–20	24–28
Lowest point voltage (per unit)	0.9128	0.9312	0.9378
Network loss (kW)	203.6	150.3	142.37

BIBLIOGRAPHY

[1] D. Shirmohammadi, H.W. Hong, Reconfiguration of electric distribution networks for resistive line losses reduction, IEEE Trans. Power Deliver. 4 (2) (1989) 1492–1498.

[2] S. Civanlar, J.J. Grainger, H.Yin, et al., Distribution feeder reconfiguration for loss reduction, IEEE Trans. Power Deliver. 3 (3) (1988) 1217–1223.

[3] L. Wang, Intelligent optimization algorithm and its application, Tsinghua University Press, Beijing, (2001).

[4] J. Zhao, Distribution network reconfiguration based on the integration of adaptive genetic algorithm and ant colony algorithm, Lanzhou University of Technology, Lanzhou, (2011).

[5] J.Z. Zhu, Optimal reconfiguration of electrical distribution network using the refined genetic algorithm, Electr. Power Syst. Res. 62 (2002) 37–42.

[6] M. Xia, Study on distribution network reconfiguration based on improved genetic algorithm, Guangxi University, Nanning, (2004).

[7] M. Hu, Y. Chen, Simulated annealing algorithm for optimal network reconfiguration of distribution system, Autom. Electr. Power Syst. 18 (2) (1994) 24–28.

[8] G. Chen, J. Li, G. Tang, Tabu search-based algorithm of distribution network reconfiguration, Proc. Chin. Soc. Electr. Eng. 22 (10) (2002) 28–33.

[9] H. Mori, Y. Ogita, A parallel tabu search based method for reconfigurations of distribution systems, IEEE Power Engineering Society Summer Meeting, IEEE, Piscataway, NJ, 2000.

[10] X. Jin, J. Zhao, Load balancing distribution network reconfiguration based on improved binary particle swarm optimization, Power Syst. Technol. 29 (23) (2005) 40–43.

CHAPTER 8

Optimization and Solution of Unit Maintenance Scheduling Models

Contents

8.1 INTRODUCTION

Unit maintenance plans are of great significance, not only from the perspective of national electricity demand, but also from the point of view of the demand coming from enterprises, in order to achieve economic benefit maximization. As such, unit damages or abnormal operation must be avoided. The power system is determined by physical conditions of the unit system, and even small changes in the physical parameters of the power system will affect the index parameters of each device, since equipment aging, or failure, will happen due to long utility time or to damages caused by nonartificial factors, making it unable to function normally or fully [1]. The failure states of equipment are divided into repairable state and irreparable state, and most of the faults are in repairable state; consequently, power equipment should be overhauled regularly without affecting the normal operation of the power system, in order to maintain the system

X. Meng and Z. Pian: Intelligent Coordinated Control of Complex Uncertain Systems for Power Distribution Network Reliability. http://dx.doi.org/10.1016/B978-0-12-849896-5.00008-8

in the best condition, or to delay the coming of the next failure [2]. The maintenance process of electric power equipment includes inspection, repair, parts update and maintenance, and so on. Although various costs, such as human and materials expenses, and inspection fees, will be generated during the maintenance process, it is necessary to arrange reasonable and appropriate unit maintenance plans to meet the stability requirements of the national electric power supply system, and the enterprise's long-term goal in economic interests.

The optimization model of the traditional generating unit maintenance plans is aimed at economy and reliability. Economy refers to building the target model from the perspective of maintenance fees, including repair cost and production expenses, and so on. The reliability model is divided into two categories, as deterministic models and random models; and deterministic models use all the feasible resources provided by the system when the unit stops operating, due to the maintenance as standard optimization, yet the random model takes into account the randomly forced shut down due to the uncertainty of loads. Constraint conditions consider the aspects of overhaul cycles, priorities and resources, and so on. With the reformation of the electric power market, the "separation of plant and grid" makes the power plant consider not only the security and stability of operation in the process of power supply, but also economic benefit maximization, while the scheduling departments mainly focus on the unification and coordination of power plant scheduling tasks; these, however, will inevitably produce certain contradictions. Therefore, it is necessary to improve the optimization model of the traditional generating unit maintenance plan.

8.2 ESTABLISHING THE OPTIMIZATION MODEL OF UNIT MAINTENANCE PLAN

8.2.1 Traditional Maintenance Scheduling Optimization Model

For this kind of multiobjective optimization problems of unit maintenance plan with multiple constraints, it is very difficult, in general, to establish a unified optimization model integrating all the influential factors, because, in the process of overhaul of plan management, some random factors that are difficult to control often appear, such as the change of electricity prices, the randomness that the unit is forced to shut down, the uncertainty of resources and the variability of residential electricity loads – all of which will affect the effects of the maintenance plan, so reliability and economy are the two factors often considered by maintenance management personnel internationally in order to make them the main targets of maintenance scheduling optimization. The following discussion will expound from the angle of objective functions and constraint conditions:

1. Objective functions

   a. The economic model. In terms of the stability and security of the power supply, the most important requirement is reliability, but, driven by the "separation of the enterprise from administration," companies are putting economy as a priority in every industrial activity, mainly aimed at achieving maximum economic benefits by combining reliability with economy, especially when the national electric power department undertakes the reform of the market system. Generally speaking, economy and reliability are conflicting. High reliability requires much more for the greatest installed capacity of the system, with higher investment cost and less economy; on the other hand, high economy demands reducing costs and a smaller installed capacity of the system, with less reliability, leading to high vulnerability of the power supply, threatening household power consumption. Therefore, a maintenance plan with high reliability may not be economic at the same time; however, this is the maximum target pursued by the enterprises in the market economy system. In conclusion, economic models are more suitable for the living mode of enterprises in the market economy system, than the one that only takes technology into account.

      The economic target is analyzed mainly from two angles, maintenance and production, and, for the repair expenses in the maintenance process, it can also be subdivided into fixed maintenance cost, and random maintenance cost. Fixed maintenance cost refers to a series of costs produced by repairing the unit alone, including various aspects, such as human, materials, as well as components and parts; random cost is the indefinable expenses of start and stop losses due to the long time of use, and the losses caused by different levels of artificial restoration, and so on.

Production expenses, in view of the production, refers to the fuel fees consumed in order to generate certain electric power, including the termination of the operation during the maintenance process, and its economic losses (interruption duration* spot prices) caused by the production shut down in the present session, and so on. Currently, some studies show that production expenses are not sensitive enough for some maintenance models, but others regard it as an influential factor that is very necessary to be taken into account.

In general, economic models study the issue from the perspectives of reducing maintenance cost, prolonging service time, and ensuring basic reliability, and so on. Here, reliability must be satisfied, so economy is often regarded as an objective function, while reliability is seen as constraints to differentiate that are finally fitting to the comprehensive evaluation model. This idea not only satisfies the security requirements of stability, but also meets the demand of enterprises for economic benefit maximization, under the market economy system, with profound practical and academic significance; therefore, it has been widely used in practical life.

b. Reliability model. The reliability model considers mainly the size of net reserve rate of the power system, during the overhaul, as the main optimization goal, with reliability as the objective function, in order to guarantee, at the same time, the stability and security of power supply. The equivalent backup method is a kind of means of optimization that regards reliability as the main optimization goal. Reliability is there to guarantee the system's stability and safety [3]. So, by referring to different factors, the reliability model can be divided into two kinds, the deterministic reliability model, and the probabilistic reliability model:

   - Deterministic reliability model. With the net reserve rate of the system as the main reference factor, the deterministic goal is to guarantee an adequate reserve rate in the process of maintenance, without affecting the power supply. Research needs to be made from the perspective of the net reserve to optimize, by referring to the greatest installed capacity and load level during the preparation of the process of maintenance, and the relationship between the real-time backup rate of the unit and the maximum net reserve rate is observed in the maintenance process, in order to bring the two back into balance as far as possible. The concept of the deterministic reliability model is very simple, and it is easy to be implemented. It is enough to arrange the maintenance in the period of maximum generation capacity, under the condition that the greatest installed capacity and load level will keep constant; so, such models have been widely used. But because of its characteristics, it is necessary to keep unchanged some influential factors such as load level, ignoring the randomness of the system, so in the period when the system loads or installed capacities make changes, the levels of reliability are also different, and the optimization result might not be the best one.

- Random reliability model. In another way, reducing the risk rate of the system is also a way to guarantee system reliability, but the risk occurs with a certain probabilistic randomness that is caused by internal factors (including the uncertainty of loads) of the system that are very difficult to control artificially, so probability analysis is needed to study this kind of random reliability models. In the whole cycle of system maintenance, this model studies the risk indicators affecting the system, such as loss of load probability, LOLP and expected energy not supplied, EENS, and so on. The minimal risk probability of the system in the cycle is considered as the reliability target, and different random reliability models are established, according to different indicators of the unit, with the same main research objective of minimal or relatively smaller risk probability in the maintenance cycle [4]. The random reliability model takes into account a variety of random factors in the system, making it more practical, fulfilling the demand of the maintenance process for randomness. However, from the data observed from the factors such as load randomness and forced shutdown of the unit, they change a lot in different time periods, leading to errors occurring in these original observation data, so optimization results cannot achieve the best results, either.

2. Constraint conditions

Constraint conditions refer to the existence of some objective factors in the unit that needs to be considered within the unit maintenance process, and while reliability is taken into account as a constraint condition; so, from the technical perspective, constraint conditions mainly involve the following aspects:

a. Constraints of the repair cycle: repair cycle constraint defines the time limit of the maintenance that must be started and finished, as required by each unit in the system.

b. Resources constraint: in maintenance scheduling optimization problems, resources constraints have been defined as all types of resources allocated by the maintenance task that should not exceed the resource capacity in the related time period.

c. Priority constraints: priority constraints reflect the relationship between the generating units in the power system; for example, the maintenance of unit 2 should normally not get started before the maintenance of unit 1.

d. Reliability constraint: reliability constraints are made in various ways, including the reserve demanded by the generating capacity during the whole planning period.

From macro perspectives, constraint conditions should also include the following contents:

1. In the case when two devices are able to be overhauled in one power outage, it must be arranged so that they will get repaired at the same time in order to avoid the losses caused by the second power outage.

2. If two devices or two lines are complementary in the power system state, they must not be overhauled at the same time. Usually, complementary state lines are designed to guarantee that there are no power outage loss of devices or lines under

maintenance or failure, so such devices or lines must not be scheduled for repairs in the same time period.

3. Some equipment in the maintenance state could be forced to shut down, so, as per the maintenance scheduling optimization model, this problem can be taken into consideration as an influential factor that, however, is not a widespread condition, and can be neglected.

4. The establishment of a system's maintenance plan needs to consider the geographical factors' effect on the maintenance plan. By arranging reasonable maintenance lines and on the premise that certain maintenance staff and funds are guaranteed, other repair expenses, such as travel expenses won't be produced, as these will increase the repair cost, affecting the economy of the maintenance plan.

5. Before establishing the maintenance plan, it is necessary to make sure which maintenance task's time is final; for example, for the maintenance tasks left over from the previous period, their maintenance time must be arranged in accordance with the previous maintenance period.

8.2.2 The Maintenance Scheduling Optimization Model Established in This Section

1. Objective functions

According to statistics, unit maintenance scheduling optimization of power plant belongs to the research field of safety operation optimization in the power system that, in fact, is a combinatorial optimization problem with multiple optimization objectives and constraints [5]. The minimum maintenance costs of all the devices (including repair costs and production losses caused by the power outage) are treated as an objective function that can be expressed as:

$$\min f(x) = \sum_{t=1}^{T}\sum_{i=1}^{N}(C_{it} + pP_{it}) + m\sum_{i=1}^{N}\left|R_{i0} - R_{i}\right| \qquad (8.1)$$

In the expression, C_{it} is the maintenance cost of device i at the time period of t; p is the coefficient of production expenses; P_{it} is the output power of device i at time interval; m is the extra cost for each change in periods of time referring to the initial stage of declaration; R_{i0} is the initial stage of declaration; R_i is the initial stage after optimization; N is the number of devices; T is the total number of time intervals.

Here, if a week is taken as a period of time, the maintenance cycles for one year will include 52 periods. According to the actual situation, a day, a month or a 10-day period can also be regarded as a period. Here, for the convenience of calculation, a week is considered as the period of time.

2. Constraint conditions

Related constraints of unit maintenance scheduling optimization problems include some common restriction factors, such as maintenance cycle, available resources, and maintenance priority [6]; and for those regarding economy as the main optimization goal, reliability should be processed as a constraint condition that can guarantee the security and stability of the power supply on the premise of ensuring economy.

Maintenance time constraints define the time limit on which maintenance must be started and finished, as required by each unit in the system, whose formula is as follows:

$$T_n = \left\{ t \in T_{plan} : e_n \le t \le l_n - \overline{d}_n + 1 \right\} \tag{8.2}$$

In the expression, T_{plan} is a discrete set of time indicators, $T_{plan} = \{1, 2, \cdots, 365\}$; e_n is the earliest time to start the maintenance allowed by the device; l_n is the latest finishing time allowed by the device; $\overline{d}_n$ is the period of time when the device needs maintenance.

In unit maintenance scheduling optimization issues, resource limits have already been defined as a condition that all types of resources allocated by the repair task should not exceed the resource capacity at the related time interval, as results from the following formula:

$$\sum_{d_n \in D} \sum_{k \in S_{n,t}} R_{n,k}^r \le \overline{R}_t^r \tag{8.3}$$

In the expression, $t \in T_{plan}$; $r \in R$; D is the set of units to be repaired, and $D = \{1, 2, \cdots, n\}$; d_n is the number of devices to be repaired; $R^r_{n,k}$ is the resource required by d_n at time interval k; $\bar{R}^r_t$ is all the related resources able to be provided at time interval k.

Priority constraints reflect the relationship between generator units in the power system; for example, the maintenance of unit number 2 should generally not be started before unit number 1, expressed as:

$$T_2 = \left\{ t \in T_{plan} : l_2 - \bar{d}_2 + 1 > t > s_1 + \bar{d}_1 - 1 \right\} \tag{8.4}$$

In the expression, $t \in T_{plan}$; s_n is the beginning time selected by d_n.

Reliability constraints refer to the reserve to guarantee the capacity demand in the whole planning period, expressed in the following formula:

$$X_{n,k} P_n - L_t \geq R_t$$
$$\sum d_n \in D \tag{8.5}$$
$$\sum k \in S_n$$

In the expression, $t \in T_{plan}$; L_t is the loads consumed at time interval t; P_n is the adjustable capacity of d_n; $X_{n,k}$ is the maintenance state of unit n at time period k, whose value is 0 or 1; R_t is the minimal backup capacity allowed.

3. The processing of constraint conditions

Unit maintenance scheduling optimization problem is a typically complex optimization issue, with multiple optimization objectives and constraint conditions. In the process of solving unit model optimization through the ant colony algorithm, the method of penalty functions is usually adopted to deal with the constraints in order to make them fit into a comprehensive objective function [7]. But the penalty function also has its own limitations. When processing constraint conditions, if they are added to the objective function in the form of penalty factors and penalty function items, the selection of penalty factor will affect the optimization results generated by the comprehensive function. If the value of the penalty factor exceeds a certain range, the optimization result may not be optimal; if its value is below a certain range, it may make the impact of constraint conditions on the objective function not important enough to show its binding function. So, in general, when determining a penalty function, many experiments are used for checking, in order to get a relatively appropriate value [8]. Here, a figure of one order of magnitude above the maintenance cost is selected as the value of the penalty factor.

For the others that cannot be fitted in by the penalty function, they will be taken into account in the process of maintenance:

a. Repair simultaneously. Devices of upper and lower levels, or the uniform loads that need to be repaired simultaneously, are scheduled for maintenance in the same period of time.

b. Don't change the maintenance time. The maintenance time of some equipment is fixed, so that, at a corresponding time interval during maintenance, they should be scheduled for examination and repairs.

c. The start time of the maintenance. For devices with a required start time for maintenance, their start and finish time of maintenance are achieved by changing the upper and lower limits of the time variable during the process of determining the objective function.

4. Optimization model

The optimization model set up in this section is based on current common regular maintenance, also called scheduling maintenance. This kind of maintenance method is based on the average life period of the device, the failure frequency and the average life cycle of lines, and so on, that have accumulated over a long period of time. Through long-term experience in repair work, maintenance personnel have basically grasped various conditions of use of the devices, so these devices could be maintained in the appropriate time intervals, avoiding the repair costs of devices under normally operating conditions. For the traditional regular maintenance plan of the generating set, their declarations of maintenance plans are able to be achieved from experience, including the unit number, power generation, scheduled period for maintenance, allowed repair cycle, the initial maintenance time for reference and maintenance costs, and so on, all having a certain reference value.

According to the previous content, the methods of regular maintenance have the following features:

a. Failures may occur at any time between the two regular maintenance periods of time, and for this kind of unexpected faults, the maintenance method can do nothing.

b. For a device at the initial stage of installation, due to various influences of environmental parameters, it needs a certain adaptation period, so, during this interval, faults will happen frequently, and therefore it needs more attention and maintenance. In addition, when the device is just about exhausted, many faults will occur due to aging. Besides the two periods, the device will experience a period of stable life cycle with a very low frequency of failure. In this case, if devices are not examined and maintained regularly, according to the actual situation, the preventative inspections cannot be carried out, while some unnecessary maintenance is undertaken, causing a series of waste of manpower and material resources. If mistakes occur in the process of maintenance, this may even cause some accidents, leading to great losses that wouldn't have happened otherwise.

c. With the continuous development of electric technology, new devices continue to emerge. Therefore, some old experience can no longer meet the needs of the new devices for the supportability of scheduling maintenance.

As such, based on the mentioned characteristics of regular maintenance, the declaration plan of periodic maintenance should refer to the declaration date, as well as make appropriate adjustments according to the actual conditions, so that it can make the final results not only satisfy the validation of experience, but also take into account a variety of inevitable factors in real–life situations. According to the characteristics of the ant colony algorithm in the optimization process, the statistics of declaration maintenance plan are rendered abstract in order to form a data space composed by virtual units. Inside the space, virtual models of the units to be repaired are established, and each model contains the

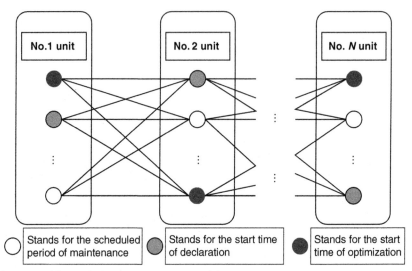

Figure 8.1 The optimized maintenance model.

time point of the maintenance-scheduled period that forms a matrix of time points in the data space. Artificial ants will search the optimum on these time points, and determine the marching direction of each step, based on state transition probability.

Thus, a declaration plan model is set up: N units correspond to the sequence of maintenance-scheduled periods allowed, and each node matches a date within the scheduled period. The nodes contain three properties, such as, power generated on that day, maintenance cost, and repair cycles. After finishing the search process, ants will output a series of node sequences based on the evaluation functions – a set of time sequence of a unit's maintenance dates that satisfy economy and reliability, as shown in Figure 8.1.

Intuitively, a unit maintenance scheduling optimization problem is a group of combinatorial problems with constraints that looks for a group of optimal initial dates of maintenance $F = \{f_1, f_2, \cdots, f_n\}$ based on the output power and maintenance cost of each unit at each time interval, under the condition that the initial dates of declaration are known, and every change in the initial period will cost a bit more. For the same units, we assume to refer to the declaration date node, and once a node is moved (a date is changed), m units of costs will be produced, and this cost is also included in the objective function as one of the influential factors.

In conclusion, the method of unit maintenance plan optimized by the improved ant colony algorithm is carried out according to the following steps:

a. According to the model diagram of the declaration plan, an artificial ant is placed at the start time of each unit's declaration, and is able to move back and forth.

b. After starting to search, ants move back and forth to determine the next target, according to state transition probability.

c. A path is formed after the ant finishes the whole set of units, also known as a time series of maintenance.

d. Determine the route cost of the path according to the objective function, that is, the maintenance cost and expenses of this maintenance time series. Confirm the local optimal value and record the information, as well as the time sequence of the route.

e. During the second iteration, ants continue to search and output the optimal maintenance time series in this time of iteration.

f. Stop searching until the maximum number of cycles is reached.

g. Compare the local optimum of each time of iterations, and output the information of the path with the global optimal value that is also the best maintenance time sequence.

8.3 THE SOLUTION OF UNIT MAINTENANCE PLAN BASED ON IMPROVED ANT COLONY ALGORITHM

As widely known, there are many methods to optimize the maintenance plan of a power plant, including the traditional mathematical programming methods, such as dynamic programming, integer programming, and mixed integer programming; and modern evolutionary algorithms, such as genetic algorithm, simulated annealing algorithm, search algorithm and ant colony algorithm, and so on, all of which have been widely used in maintenance plan optimization. Compared with other optimization algorithms, the ant colony algorithm has higher calculating rates and quality, such as the traveling salesman problem, in solving a series of combinatorial problems of optimization projects. But in the process of application, the short-comings of the ant colony algorithm have appeared gradually, such as poor a global search ability, long search time, and the fact that it is easy to fall into local optimum. This section is intended to adopt fuzzy control rules in order to improve the ant colony algorithm, and apply the improved algorithm to the established optimization model of the power station maintenance plan.

8.3.1 Updating the Rules of Pheromones

It is easy for stagnation problems to appear in traditional ant colony algorithms, and for the algorithm to fall into local optimum, mainly because of the updating rules of pheromones. When the number of pheromones is too small, it is easy for the algorithm to experience a stagnation phenomenon; yet when the number of pheromones is too large, ants will concentrate on one path, falling into the state of local optimum [9]. In terms of this problem, equation (2.10) is improved to propose an additional updating rule of pheromones.

$$\tau_{ij}(t+1) = \begin{cases} \tau_{min} & \tau_{ij}(t+1) < \tau_{min} \\ \tau_{ij}(t+1) & \tau_{min} < \tau_{ij}(t+1) < \tau_{max} \\ \tau_{max} & \tau_{ij}(t+1) > \tau_{max} \end{cases} \quad (8.6)$$

In this way, the quantity of pheromones between nodes can be controlled in a certain range, solving the problems of stagnation and falling into local optimum.

Meanwhile, from the above description it can be seen that $g(t)$ and ρ are two important parameters affecting pheromone update, of which $g(t)$ mainly affects the global convergence rate, and if its value is too large, results will fall into local optimum; if its value is too small, the research ability of the algorithm is improved a bit, but the research range and time will reduce the convergence rate. The main role of ρ is to control the global searching ability, affecting the convergence rate indirectly; and if its value is too large, the volatile rate of phero-mone is so fast that the searching process becomes more random and globalized, but the convergence rate will also be reduced because of it. If its value is too small, a phenomenon will occur such as some paths are chosen twice in the optimization process [10]. Therefore, the valuing of $g(t)$ and ρ is of great importance to the search results, so fuzzy control rules will be adopted in order to improve it.

8.3.2 Improve the Ant Colony Algorithm by Fuzzy Control Rules

1. Fuzzy logic theory

In the mid-1960s, Professor L. Zadeh published a paper on the theory of fuzzy sets, officially creating a precedent of fuzzy theory. Therefore Professor Zadeh is also known as the "father of the fuzzy theory." Later research on fuzzy theory has increased steadily, and has also been applied gradually into practice [11,12].

Professor Zadeh proposed that, usually, there are no clear boundaries between the concepts or facts in nature, so, in order to distinguish them, he put forward two concepts:

a. Ordinary set. The ordinary set has clear divisions or boundaries, and the characteristics of objects in the set are clearly visible. For example, for the integer set ranged between 3 and 8, $M = \{\text{integer } r \mid 3 < r < 8\}$, and the values of r are a few clearly figures visible, so it is an ordinary set.

b. Fuzzy set. The fuzzy set is a set without clear boundaries, and objects in the set are characterized as qualitative, rather than quantitative. For example, for an integer set close to 8, the term "close" cannot be determined by a precise function or rule, and only the degree of the object's membership to the set can be judged, called membership grade [13], which is a real number defined between [0,1]. For the "integer set close to 8," the membership grade can be defined as the membership grade of 7.9 and 8.1, that is, 0.99, while for 5 and 11 it is defined as 0.7. This continuous multivalued logic is called fuzzy logic.

In terms of the membership grade [0,1], although its range of value is the same as that of probability in probability theory, the meanings are different, and they must be clearly distinguished.

Objects discussed in probability theory must be clear, but the probabilities of their occurrences are not clear, such as "the probability of individuals suffering from heart disease inside a specified number of groups" can be achieved by the probability statistical method.

However, objects described by membership grade are not clear, and their occurrence can be definite or uncertain, for example "cold" and "very cold" will definitely appear in life, but there is no definite statistical method to divide them.

2. Fuzzy control rules of the improved ant colony algorithm

With regard to the process of some complex uncertain events, it is hard to control them because of their nonlinear, time varying, and random control process that can be solved in two ways: one is to establish a mathematical model to control it – but for some uncertain events, it is difficult to establish a precise mathematical model to describe them, and once the control process is beyond the scope of model's application, it becomes useless; another way is to establish a set of control commands taking into account various cases based on historical experience that searches for the appropriate control strategy according to established situations [14–16], and this kind of control command sets are called fuzzy control rules. Because of its characteristics, fuzzy control is very suitable for such a control process – also a field where fuzzy logic is most widely used. Rules functions of the computer are able to control the process of events; however, they can only achieve two results, true or false, because of its characteristics [17]. But because the membership grade of the fuzzy control rule is a continuous range from which the value is selected, the accuracy is much more guaranteed.

The following example illustrates the application of fuzzy control rules. Changes throughout the year mainly have two influential variables, temperatures and precipitation (including rain and snow); these will produce a set of changes within a year. In Beijing, for example, the temperature changes following the four seasons, with the changes set containing "warm," "hot," "cool," and "cold," and the changes set of precipitation includes "little," "many," "less," and "minimum," while their output variable set is "spring," "summer," "autumn," and "winter." Thus, the following control rules are output based on the established virtual control model:

If (temperature is "warm") and (precipitation is "little") then (this season is "Spring");

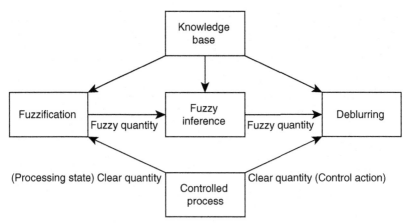

Figure 8.2 The process of typical fuzzy control.

If (temperature is "hot") and (precipitation is "many") then (this season is "Summer");

If (temperature is "cool") and (precipitation is "less") then (this season is "Autumn");

If (temperature is "cold") and (precipitation is "minimum") then (this season is "Winter").

In the fuzzy control system shown in Figure 8.2, the fuzzification interface converts the process state generated during the control process into fuzzy quantity, usually expressed through fuzzy sets. After the conversion, the generated fuzzy quantity carries out fuzzy inference through the core steps of fuzzy control, and the fuzzy inference simulates human thinking, mainly based on the implication relations of fuzzy logic, also known as fuzzy rules. Deblurring is a process corresponding to fuzzification, reducing the fuzzy quantity generated by fuzzy inference to the control behaviors understood by human beings. The knowledge base contains the knowledge and experience of areas related to the control process and goals to be achieved by the control process, generally composed by database and fuzzy control rules [18].

It can be seen from the previous analysis that it is easy for traditional ant colony algorithms to fall into stagnation and local optimum, mainly because of the updating rules of pheromones. When the pheromone is too small, the algorithm is prone to stagnation; yet when it is too large, ants will concentrate on one path, falling into the local optimum state. To solve this problem, the fuzzy logic reasoning system is adopted in order to define fuzzy control rules according to the randomness and fuzziness of two parameters, bringing the valuing of $g(t)$ and ρ under control. Here, the values of $g(t)$ and ρ are obtained from experiments with $g(t) \in [100, 400]$ $\rho \in [0.1, 0.4]$.

Define the rules as follows:

if the continuous S generations haven't evolved a new result, then select a smaller $g(t)$ and a bigger ρ;

if a new result shows up,

then the values of ρ and $g(t)$ remain unchanged.

Thus, the values of ρ and $g(t)$ will adjust according to the state change of the ants' searching process, and when the local optimum occurs in the algorithm, on the one hand reduce $g(t)$, and from equation (2.10), obtain:

$$\Delta\tau_{ij}(t) = \begin{cases} g(t)/C & \text{there are ants passing by} \\ 0 & \text{there are no ants passing by} \end{cases}$$

As such, the increments of pheromones will drop a bit, and thus the pheromone value of some node will tend to fall. One the other hand, select a bigger value of ρ and from equation (2.9), achieve:

$$\tau_{ij}(t+1) = (1-\rho)\tau_{ij}(t) + \Delta\tau_{ij}(t)$$

Therefore, the volatility of pheromones will get strengthened, also leading to a drop in the value of the node pheromone. Through these adjustments, the pheromone's guiding roles in the artificial ant optimization process will be weakened, while the impact of other factors, such as distance, will be strengthened on the ant's state transition probabilities so, in this way, the ant's searching space is expanded, finally escaping local optimum issues.

The process of adopting fuzzy control rules to improve the ant colony algorithm is shown in Figure 8.3.

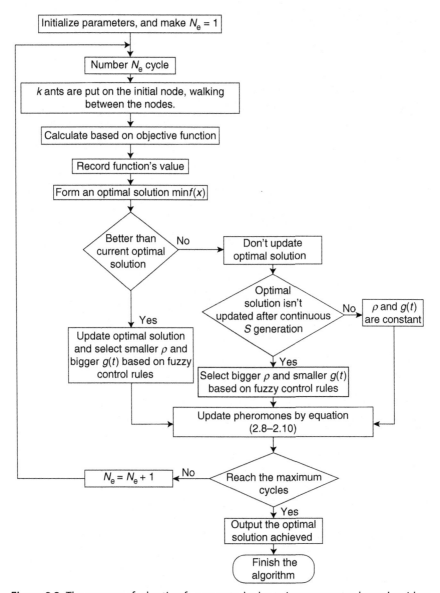

Figure 8.3 The process of adopting fuzzy control rules to improve ant colony algorithm.

8.4 EXAMPLE ANALYSIS

8.4.1 Model Validation

8.4.1.1 Validation

In order to verify the effectiveness of the model described in this chapter, the results of the optimization model mentioned in reference [19] will be compared with those of the model established in this chapter. Take the overhaul plans of 10 units as an example, where the scheduled maintenance period is defined as the time period between the earliest start time and the latest start time allowed by the units; the initial period of declaration is the start time of the optimal maintenance plan achieved from historical experience by the scheduling department; the maintenance cycle is the time to repair allocated for each unit by the scheduling department, usually without modifications; maintenance costs are the repair fees for each unit; the generated power is used to calculate the economic losses caused by the shutdown. Here, we assume to refer to the target of minimum function value after optimization, and every change in the original declaration plan for one maintenance cycle (change refers to being advanced or postponed) will add 10 Yuan to the maintenance costs. For convenience's sake, we assume that the case that units are forced to stop operation randomly will be not taken into account. The declaration maintenance plans of 10 units are shown in Table 8.1.

Table 8.1 The declaration of maintenance plan of 10 units

Number	Capacity (MW)	Scheduled maintenance period (week)	Initial stage of declaration (week)	Maintenance cycle (week)	Maintenance cost (Yuan)
1	250	1,8	2	6	100
2	300	3,12	6	6	150
3	150	2,6	3	2	100
4	300	5,18	8	5	150
5	300	11,19	12	5	200
6	200	13,18	15	3	200
7	200	8,14	10	3	150
8	150	15,21	17	3	150
9	200	12,20	14	5	150
10	200	10,16	12	3	100

Table 8.2 Validation of the maintenance scheduling models of 10 units

Model optimization results of Reference [19]		Model optimization results established in this chapter	
Number	**Start time and cycles of maintenance**	**Number**	**Start time and cycles of maintenance**
2	5(6)	2	4(6)
5	13(5)	6	14(5)
8	16(3)	7	11(3)
		10	13(5)
Total expenses (Yuan)	75624	Total expenses (Yuan)	74853
Time (s)	32	Time (s)	37

The optimization algorithm adopts the traditional basic ant colony algorithm, whose parameters are set as $m = 50$, $\alpha = 4$, $\beta = 1$, $\rho = [0.1, 0.4]$, $g(t) = [100, 400]$, $g_{max} = 200$, $c_1 = c_2 = c_3 = 1000$, and results are shown in Table 8.2.

The experimental results show that, when optimized by the optimization model established in this chapter, the maintenance times of unit 2, 6, 7, and 10 are adjusted, and the total expenses are reduced from 75,624 Yuan to 74,853 Yuan, with very little changes in optimization time. Therefore, the model is proved to be able to achieve a better maintenance plan, suitable for the economy requirements of the power enterprises on the premise of ensuring its reliability under the same conditions.

8.4.1.2 Robustness Verification

Regarding the traditional ant colony algorithm as the basic optimization algorithm, the robustness of the model established in this chapter is verified from two angles, respectively, the different number of units, and the different groups of ant colony algorithm.

1. Different number of units. Take for optimization the maintenance planning data of 20 units, whose declaration of maintenance plans is shown in Table 8.3.

   The optimization algorithm adopts the traditional basic ant colony algorithm, whose parameters are set as $m = 50, \alpha = 4, \beta = 1$,

Table 8.3 The declaration of maintenance plan of 20 units

Number	Generated power (MW)	Scheduled maintenance period (week)	Initial stage of declaration (week)	Maintenance cycle (week)	Maintenance cost (Yuan)
1	150	1,9	2	6	200
2	200	2,11	4	5	250
3	250	3,6	4	2	300
4	100	4,17	6	6	150
5	100	10,15	11	3	100
6	300	11,20	13	6	100
7	100	6,12	8	3	250
8	250	13,19	14	3	150
9	300	14,21	16	4	150
10	300	11,22	12	5	200
11	250	12,23	15	7	100
12	150	13,16	14	2	100
13	100	9,14	10	3	300
14	200	15,24	16	7	200
15	200	22,25	23	2	250
16	300	16,21	17	3	100
17	250	23,28	24	4	250
18	200	20,22	20	1	100
19	300	19,27	21	5	200
20	100	26,29	27	2	200

$\rho = [0.1, 0.4]$, $g(t) = [100, 400]$, $g_{max} = 200$, $c_1 = c_2 = c_3 = 1000$, and results are shown in Table 8.4.

2. Different groups of ant colony algorithm. Take the declaration of maintenance plans of the units in Table 8.1 as optimization goals. The groups in the ant colony algorithm $m = 100$ and other parameters are as the same: $\alpha = 4$, $\beta = 1$, $\rho = [0.1, 0.4]$, $g(t) = [100, 400]$, $g_{max} = 200$, $c_1 = c_2 = c_3 = 1000$, and their verification results are shown in Table 8.5.

8.4.2 Verification of Improved Algorithm

To validate the improved algorithm, the grid world is regarded as the simulation environment in order to calculate the shortest

Table 8.4 The optimization results of the maintenance scheduling model of 20 units

Model optimization results of Reference [19]		Model optimization results established in this chapter	
Number	Start time of maintenance	Number	Start time of maintenance
4	7	2	5
6	12	6	14
7	7	7	9
11	13	14	17
13	11	15	22
18	21	20	26
19	20		
Total expenses (Yuan)	164526	Total expenses (Yuan)	149784
Time (s)	44	Time (s)	41

distance between two nodes on the opposite angles of the grid world, by operating two algorithms respectively: basic ant colony, and improved ant colony algorithms. The parameters of the basic ant colony algorithm are: $m = 50$, $\alpha = 4$, $\beta = 1$, $g_{max} = 200$. The parameters of the improved algorithm are set as: $\alpha = 4$, $\beta = 1$, $\rho = [0.1, 0.4]$, $g(t) = [100, 400]$, $g_{max} = 200$, all of which are achieved from experiments.

Table 8.5 Validation of the maintenance scheduling models of 20 units

Model optimization results of Reference [19]		Model optimization results established in this chapter	
Number	Start time of maintenance	Number	Start time of maintenance
3	4	4	9
5	13	7	11
6	14	9	15
7	9		
Total expenses (Yuan)	74485	Total expenses (Yuan)	74842
Time (s)	38	Time (s)	31

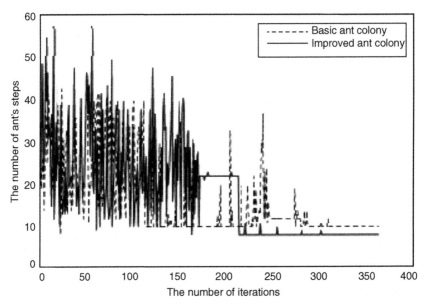

Figure 8.4 Simulation and verification results of the improved algorithm.

The simulation and verification results of the improved algorithm are shown in Figure 8.4, with searching abilities and convergence rates all improved.

BIBLIOGRAPHY

[1] J. Wang, X. Wang, C. Feng, et al., Generator unit maintenance plan based on market fairness, Autom. Electr. Power Syst. 30 (20) (2006) 15–19.
[2] M. Ding, Y. Feng, Study on reliability and economy of generator unit maintenance plan, Electr. Power 34 (7) (2001) 22–25.
[3] J. Yellen, T.M. Al-Khamis, S. Vermuri, et al., A decomposition approach to unit maintenance scheduling, IEEE Trans. Power Syst. 7 (2) (1992) 726–733.
[4] Y. Feng, Maintenance Scheduling of Power Generation and Transmission Based on Generalized Benders Decomposition in Power Market, Hefei University of Technology, Hefei, (2002).
[5] K. Kawahara, A proposal of a supporting expert system for outage planning of electric power facilities retaining high power supply reliability, IEEE Trans. Power Syst. 13 (4) (1998) 1453–1465.
[6] Y. Liu, Theoretical Research on Ant Colony Algorithm and its Application, Zhejiang University, Hangzhou, (2007).
[7] S. Zhao, L. Wang, Application of improved ant colony algorithm in distribution network planning, Power Syst. Prot. Control 38 (24) (2010) 62–64.

[8]   W. Sun, W. Shang, D. Niu, Application of improved ant colony optimization in distribution network planning, Power Syst. Technol. 30 (15) (2006) 85–88.

[9]   L. Munoz Moro, A. Ramos, Goal programming approach to maintenance scheduling of generating units in large scale power systems, IEEE Trans. Power Syst. 14 (3) (1999) 1021–1028.

[10] T. Satoh, K. Nara, Maintenance scheduling by using simulated annealing method, IEEE Trans. Power Syst. 1 (6) (1991) 850–857.

[11] K. Suresh, N. Kumarappan. Combined genetic algorithm and simulated annealing for preventive unit maintenance scheduling in power system. IEEE: Power Engineering Society General Meeting, 2007:18–22.

[12] S. Chen, P. Yang, Y. Zhou, et al. Application of genetic and simulated annealing algorithm in generating unit maintenance scheduling, Autom. Electr. Power Syst. 22 (7) (1998) 44–46.

[13] Y. Wang, E. Handschin. Unit maintenance scheduling in open systems using genetic algorithm. IEEE: Transmission and Distribution Conference, 1999, 1(1):334–339.

[14] R.C. Leou. A new method for unit maintenance scheduling based on genetic algorithm. IEEE: Power Engineering Society General Meeting, 2007:246–251.

[15] L. Wang, Intelligent Optimization Algorithm and its Application, Tsinghua University Press, Beijing, (2001).

[16] C.A. Koay, D. Srinivasan. Particle swarm optimization-based approach for generator maintenance scheduling. IEEE: Swarm Intelligence Symposium, 2008:167–173.

[17] Y.S. Park, J.H. Kim, J.H. Park, et al. Generating unit maintenance scheduling using hybrid PSO algorithm, IEEE Intell. Syst. Appl. Power Syst. (2007) 1–6.

[18] H. Tajima, J. Sugimoto, T. Funabashi et al. Auction price-based thermal unit maintenance scheduling by reactive TABU search. IEEE: Universities Power Engineering Conference, 2008:421–425.

[19] K.F. Wai, A.R. Simpson, H.R. Maier, S. Stolp, Ant colony optimization for power plant maintenance scheduling optimization - a five-station hydropower system, Electr. Power Syst. Res. (2007) 434–436.

INDEX

Printed in the United States
By Bookmasters